Modeling Software with Finite State Machines

Machines

A Practical Approach

Other Auerbach Publications in Software Development, Software Engineering, and Project Management

AUERBACH PUBLICATIONS

www.auerbach-publications.com
To Order Call: 1-800-272-7737 • Fax: 1-800-374-3401
E-mail: orders@crcpress.com

Modeling Software with Finite State Machines

A Practical Approach

Ferdinand Wagner
Ruedi Schmuki
Thomas Wagner
Peter Wolstenholme

CRC Press
Taylor & Francis Group
Boca Raton London New York

CRC Press is an imprint of the
Taylor & Francis Group, an **informa** business
AN AUERBACH BOOK

CRC Press
Taylor & Francis Group
6000 Broken Sound Parkway NW, Suite 300
Boca Raton, FL 33487-2742

First issued in paperback 2019

ISBN-13: 978-0-8493-8086-0 (hbk)
ISBN-13: 978-0-367-39086-0 (pbk)

Library of Congress Card Number 2005035224

Library of Congress Cataloging-in-Publication Data

Modeling software with finite state machines : a practical approach / Ferdinand Wagner ... [et al.].
        p. cm.
    Includes bibliographical references and index.
    ISBN-13: 978-0-8493-8086-0 (0-8493-8086-3 : alk. paper)
    1. Computer software--Development. 2. Machine theory. I. Wagner, Ferdinand.

QA76.76.D47M625 2006
005.1--dc22   2005035224

Preface

This book discusses a topic that is among the central questions of software development. Therefore, we must position ourselves in that area to justify our right to express our opinion on that topic. Saying "we" implies at least one person in the co-author group. We have worked for several years in software development using various languages and development environments. We did this in large, medium, and small companies, as well as individually. We took part in projects in different positions as programmers, project leaders, managers, or consultants. We know software from the university perspective as a scientific and teaching subject. Hence, we have experienced software from several perspectives. We regularly read books and journals ranging from IEEE journals to very simple marketing-controlled papers. This experience allows us to have a well-established opinion, which is independent of any pressure of specific professional circumstances. For instance, I have a rather bad opinion about BASIC but some co-authors did a good job with Visual Basic and see it in a better light than I; thus, we managed to find a compromise view about that issue.

The book is organized in three parts. Two parts represent a specific view on software development encouraging the use of state machines in software design and implementation. The Vfsm method and the State-WORKS tools, which are based on that method, are not simply concepts. They have been tried over several years in many projects, proving their sense and usefulness. The reader may agree with it or not, but the discussion can be led only on a technical basis.

The first part may be the most controversial because it represents our view about software, which is per se a complex issue of not only technical character. Therefore, I explain in the first sentences of the Introduction our background. We try also to limit our opinions to technical matters.

The Vfsm concept discussed in the book is application oriented in comparison to the code-oriented approach that is typical for nearly all software development methods. There have always been trials to replace a classical software development by approaches that are more appropriate for inexperienced or occasional programmers. Interpretations of BASIC programs, PLC programming, and the Smalltalk development environment are all examples of software developments that do not require typical compilation, linking, and code debugging. Those approaches can be called "doing by trial" and are relatively safe as each written line can be tested immediately — at least for its syntactical correctness. This creates an illusion about the ease of programming. Of course, producing syntactically correct code lines easily does not mean that we are writing anything sensible. The known approaches of that kind do not provide any methods that would allow large software projects to be developed successfully.

The essential concept of StateWORKS is to use a ready-made program for realization of the most fragile part of software: the behavior. The work of the developer is then concentrated on a specification of the behavior. The input/output interface as well as any user interface must be programmed conventionally. The point is that in replacing coding of an application's behavior by a specification, we simplify the entire software development as the ready-made execution part is a frame that determines much of the software design. Effectively, many decisions about the software design are automatically made, as enforced by the execution program.

To create such a development and execution environment, several problems had to be solved. Behavior is the most complex part of software and therefore the source of most obscure errors. To replace coding based on continuous testing of conditions (if-then-else and case statements) we needed a new method and also an environment that allows for its implementation. The method used by StateWORKS is well known: modeling with state machines. To be successful, the method has also had to solve the problems of designing systems of state machines. The virtual environment is the base of a development and execution environment. The intimate meaning of those few sentences above are discussed in the book.

We must stress that the method proposed is not merely another way of programming finite state machines, embedded in other software, for certain applications — or parts of applications — which very obviously could use them. We are proposing a major re-think about the way to design software, in which most of the usual "control-flow" coding is avoided, and the state machine concept is pushed much farther than before, so as to take a predominant part in the design process.

We have written this book about a topic that has occupied us for several years. All co-authors contributed, although in different ways. F. Wagner, who invented the Vfsm concept, wrote most of the text in the book.

R. Schmuki's achievement has been in the design and implementation of the RTDB, which is the heart of the StateWORKS runtime system. P. Wolstenholme's deep understanding of control system problems was a constant inspiration to improve the tools and the text. T. Wagner introduced some fresh air in the established environment forcing us to re-think several "obvious" concepts, which led to several improvements of the tools and text.

The book uses "we" instead of an impersonal form. In rare cases, "I" is used to express my personal view: though it is my (F. Wagner's) view, the co-authors obviously support it or at least have not protested too strongly.

F. Wagner

Source Code

You can download StateWORKS development tools LE edition from www.stateworks.com. As a buyer of the book you are entitled to free registration of the software, which can then be used without a time limit. To register.

The software package contains the source code of all examples in the book and the Appendices, including those printed in the book and extra appendices for which we had no room.

Document Conventions

The book uses the following typographic conventions.

Example of conventions	Description
Vfsm	The name of the Virtual finite state machine
VFSM	The name of the Vfsm RTDB object
State_1	States are in *italics*
Timer_OVER Start_Timer	Values (Control values and Actions) are in `courier`
Format	Names when used (defined) for the first time are written in bold (these are generally cited in the Index)
StartingPressure	All words of formal names begin with a capital letter
|	Boolean OR operator
&	Boolean AND operator
Boxed text	Important text

Trademarks, Registered Marks

StateWORKS
Beckhoff
Microsoft
MS Windows
MS Word
MS Excel
MS Access
Visual Studio
Statechart
QuatroPro
MySQL
Word Perfect
Eclipse
eXtreme

About the Authors

Ferdinand Wagner holds the degree of Ph.D. in electronic digital circuits from Technical University in Gliwice, Poland. He was professor of electronics and computer science at Polish and German universities, and was an invited professor at CERN, Geneva and Swiss Federal Institute of Technology in Lausanne, Switzerland. He worked at several industrial companies in Germany and Liechtenstein developing software for control systems. He served as a consultant to AT&T, Naperville, IL, USA and to industrial institutions in Europe. He is author of several books and papers on computer hardware and software design.

Ruedi Schmuki is a project leader software engineer at Unaxis, a Swiss semiconductor equipment manufacturer. Previously, he was involved with the development of hardware and software for Contraves, based in Zürich, Switzerland. He graduated in electronic engineering and holds a postgraduate degree in system theory from the University of Applied Sciences of Technology Buchs, also in Switzerland.

Thomas Wagner is a member of the design team in mobile communication at Nortel Networks Germany. He develops services and tools for Nortel's Intelligent Network platform. He has an M.S. degree in electronic science from the University of Karlsruhe, Germany.

Peter Wolstenholme graduated with a B.Sc. from Manchester University in 1954 and stayed on for a research M.Sc. degree relating to computer peripherals. Then, at Ferranti, he worked on guided-weapon and radar systems integration, and later took project responsibility for the Argus 200 and Argus 400 control computers — the latter was the first I.C.-based computer on the commercial market. He moved to CERN, Geneva, in 1967, initially to manage a computer-control project for the Intersecting Storage Rings, and later on arranging for the introduction of fiber-

optic communication, and for extension of the CERN internal networks to cover the large site. Since 1990 he has been designing hardware and software for various small companies. He has always been an enthusiast for design methods that ensure that the designer knows the designs will work correctly, in all circumstances, so that testing becomes a formality rather than a debugging process.

Table of Contents

Part III StateWORKS: Principles and Practice

Part I

THE PROBLEMS
OF SOFTWARE

Chapter 1

Evolution of Software Development

Introduction

The book is about state machines as they should be understood and used in software. So, why have we written a few chapters about general software issues? The reason is quite simple: we introduce a new specification method based on a state machine concept. We do this in an environment more flooded with ideas, concepts, and methods than any other branch of technology or science. We do not wish to present something that is just another new idea about software development methods; what we describe is a coherent and well-tested framework for generating reliable software for even the most complex and difficult situations. We propose to show that the established practice of using a specification as a basis for coding is wrong. We also propose to show why software needs the Vfsm concept and give reasons for sponsoring the idea of specifying behavior, and execute results of the specification. Summarizing, we propose to present arguments for reading the immediately following parts of the book, which contain the material closely relevant to the topic.

The first title of this Part I was to have been "The Software Crisis." Thinking longer about it, we decided to defuse the heading. As far as we can go into the past we have always heard about the software crisis. A flourishing branch of industry cannot be in a crisis for 30 years. Obviously, it is an inherently normal state for that activity and we just have to live with it. "To live with it" does not mean to accept and to consider it as something "given by God" that cannot be changed. Especially, we know that it is a result of human activity, and as such it can obviously be changed by us. Furthermore, there are commercial reasons for change, primarily the exorbitant cost of software development.

It is much easier to formulate the goal than to implement it. We have chosen for this section a very personal, but we think a fair way. Instead of trying to formulate a very deep analysis supported by scientific methods and evaluations we simply present here our knowledge and opinions about the topic. A more thorough analysis would have required a larger team and resources, and eventually it would exceed the true goal and topic of the book.

The view of software as presented in the few chapters of the book is a view of persons who have spent many years developing software and observing it from several positions: programmer, software project leader, software manager, and last but not least academic teacher. Even if some thoughts appear not well founded, they represent a point of view that is not rare or unusual — it is simply the view of many people inside the software industry. The public official façade presented by many companies is completely false — the true face of software development, the methods really used, and the various management styles can only be seen from inside the company, when taking part in the projects.

In Part I we avoid in most cases formal definitions or very detailed explanations of issues discussed as we assume that a reader is familiar with the topics. This is in contrast to the later parts, which either present new approaches or carefully define things as we come to see that we need them. Therefore, even such an old and well-known idea as a finite state machine will be given its adequate definition in Part II. In other words, the first part is informal and avoids pedantry; the following parts are formal where we try to be as precise and clear as possible. In the first part we do not try to teach a reader anything; we just express our opinions. In the following parts we present a formal method and its implementation.

Programming Languages

Software is inherently linked with programming languages. For people outside the software industry the two are the same.

A programming language is a formal language defined by its syntax and semantics. The programming language is used to write programs that are carried out by computers. There are hundreds, perhaps thousands of programming languages, most of them irrelevant. The overwhelmingly majority of programs are written in a few dominant programming languages.

A Little History

Programming began with writing programs into a computer memory in a form of octal or hexadecimal numbers, which represented the binary-coded program instructions (machine code). For example, 0272 12601 is an indexed indirect jump for one old computer, and some programmers were happy to work with such coding. Soon the numbers were replaced by instruction names — the assembler was born. Assembler notations, just like the underlying binary code, are closely linked with the processors. So, each processor has its own assembler language. There were some efforts made to introduce a common assembler notation discussed, for example, by Baldwin.[1] One of the implementations, CALM (Common Assembly Language Notation), was presented by Nicoud et al.[2] If successful, it would have broken the bindings between the processor and the assembler. Those efforts failed and resulted only in some short-lived and locally limited implementations.

This direct fiddling with bits and bytes while programming was very time-consuming although the programs themselves were effective. In the sixties, the first general programming languages appeared that were application oriented — the direct link between a processor and a program has been removed. A given program could have been carried out on any computer. To realize it, translation programs (essentially a compiler and a linker), which converted the program into processor specific instructions, were used. The most important early programming languages were Fortran and COBOL, which were not grounded on any solid design philosophy because the theoretical basis of the emerging software practice was rather thin in those days (we should not forget that only 40 years ago relatively obvious but important observations were being published; e.g., that overuse of the *goto* statement leads to disastrous code as shown by Dijkstra[3]). Attempts to introduce languages based on some theoretical presumptions like Algol or later Ada failed to have wide impact, perhaps on account of the design-by-committee problem, although a certain number of valuable concepts were established.

The lessons of the use of the early and intuitively defined languages led quickly to the idea of structured analysis and design. The best realization of that idea was Pascal, which was *the* programming language in the 1970s to 1980s. For reasons that, from a technical point of view, are hard to understand Pascal was replaced by C, and Modula 2 was ignored. The effect of the C language could be described by a remark: "hearing about a major software catastrophe I can be sure that it has been written in C." Anyway, that language, with its descendant C++, seems at the time of writing to have won the battle for supremacy.

BASIC Catastrophe

Speaking about software catastrophes, we should mention BASIC. There are no known large software catastrophes linked with BASIC as nobody has tried to write large programs with it, but in a way it is a disaster, when considering its bad influence on programming style. BASIC was introduced as a teaching language for students of programming, or perhaps we should say coding, with an apparent advantage of interactivity, as when first introduced in the sixties it was interpreted rather than compiled. This encouraged users to just start coding, and to fight their way through a mass of errors and misunderstandings until they seemed to have a successful program. BASIC was never a favorite language for experienced programmers. Dijkstra pointed out once that a person who was taught programming using BASIC would never be a good programmer. The statement should not be interpreted directly (we are taught many stupidities in our youth but as adults most of us manage to overcome the ballast) but it describes well the serious programmers' opinion of BASIC. That language encountered an astonishing evolution, taking over several proven ideas from object-oriented languages, and a modern version, Microsoft Visual Basic, stays popular. In fact, many millions of useful small-scale projects have been implemented with Visual Basic, which provides superb facilities for a certain class of software running in the limited environment of the Windows desktop. In spite of that, C++ programmers cannot understand why they should ever use BASIC when they can do the same things in C++ better and more elegantly. Visual Basic has some serious limitations that arise if we want to do something serious. We are enthusiastic if creating nice pictures on the screen, but leaving the graphics behind we fall flat on our face very soon.

C++

We have already mentioned C++, which is the language of the last decade. It is the flagship of the idea of object-oriented programming, which evolved in the late eighties. That paradigm was the last great step in the direction of better programming languages. The success of C++ is also difficult to explain; probably its evolution from the very popular C played the decisive role. There are languages that better implement the idea of the object-oriented paradigm, especially Smalltalk, where a programmer lives truly in an object world. This is in contrast to C++, where a programmer is still confronted with the plagues of the C language, like pointers, global variables and functions, garbage collection, exception handling, etc. There are of course several other object-oriented languages, the most interesting being Java or C#. The variety of languages fulfills the goal of having some advantages over other languages from a certain point of view, in a certain environment, but we cannot say that any of the languages is in principle better than others. As we pointed out above, the dominance of C++ does not mean that it is the "best" programming language from a theoretical viewpoint. Languages defined on a sound theoretical basis have never gained wide acceptance (see the already mentioned Algol or Ada).

PLC

Speaking about programming languages we should not forget PLC (Programmable Logic Controller) programming methods. The way PLCs are programmed has never excited programmers, especially the true PLC programming in the form of ladder diagrams or function plans. In any case, PLC programming is successful and often used. The tight links between the PLC hardware and programming have not allowed any other way of doing it, which explains the success story.

The PLC world is a very separate world. It is so isolated from the mainstream of software development that some programmers do not consider it as true programming. The separation has stemmed from the close linkage to the hardware, i.e., to the Programmable Logic Controllers and the relay logic that preceded them. The application domain for the PLC is well defined and limited to industrial control, where the hardware is first needed as a data acquisition system, which is the interface between the controlled production system and the computer. For several years PLC hardware used its own execution units and for that proprietary hardware each company developed its own software development environment. After the proprietary computing systems had been replaced by open hardware solutions symbolized by PC, the PLC world delayed any changes favored by the marketing situation in that field: a few big companies shared and controlled the market. Eventually they were forced to open the PLC to nonproprietary solutions — standards for programming languages having been introduced (e.g., IEC 61131-3). Those standards are taken seriously in Europe, although the U.S. companies are delaying acceptance, still stuck on ladder diagrams, which differ slightly from one manufacturer to another one. Standards in turn have allowed the use of the PC as a control unit.

From a technical point of view, PLC programming methods (ladder diagram, structured text, function block diagram, instruction list, and sequential function chart) do not have any advantages over other programming languages, but the nature of a PLC language is such that developers are able to concentrate on the application, rather than having to spend half their time struggling to avoid or to correct programming-syntax problems. The main sense of this approach is the large potential represented by people who know the industrial control problems and can program in PLC languages and the good development tools provided by PLC manufacturers. The power of PLC manufacturers plays a decisive role here, as well. It is rather difficult to convince, e.g., a C, C++, or Java programmer to change voluntarily to PLC, but many firms have, nevertheless, chosen to adopt them after having suffered with conventional programming in the days of the mini-computer, in the same way that some firms chose Visual Basic rather than C or C++, so as to concentrate better on the application and have fewer obscure bugs in the final product. We present detailed criticisms of the IEC 61131-3 methods in Appendix E: *Going Beyond the Limitations of IEC 61131-3*.

Script or Macro Notations

A very special type of programming language is script languages, also called command or macro languages. In principle, for any job we may invent a special language. If we can, we do it. So, of course, any operating system has its script language: the IBM mainframe JCL, DOS "bat" (very poor), UNIX shell programming (quite good but unfortunately every UNIX variant has a slightly different one). Many application programs have one — called a macro language: editors (WordPerfect, Word, OpenOffice,...), databases (Access, MySQL,...), Spreadsheets (Excel, QuatroPro,...), Internet applications (Browsers,...). We had better stop the list — effectively it could contain all application programs. There were attempts to introduce BASIC as a common macro language at least for MS Windows applications, with little success.

There is a hope that eventually XML notation will become a lingua franca of the computer world. Interestingly, XML is not a programming language but just a data description language and it can be used only with an execution environment that gives the data the true interpretation. In any case, it seems that it could solve the "Tower of Babel" mess existing in script, command, and macro languages.

There Are Many Languages

In such a way we reach the end of this short history of programming languages. Readers who program at the moment in Prolog or Lisp, for example, could ask why their programming language has been omitted. There are several languages whose only reason for existence is that they have been developed for a very specific application domain and outside that domain they are just useless. The list of such specific applications and corresponding languages is long. For instance, many years ago I tried to program in APL and I was enthusiastic about the compactness of the language — for some mathematical exercises it was a perfect tool. But it was also obvious that that language is just a toy that will never achieve any importance.

As a final comment, we suggest that most programming languages do not "scale" very well. They are easy to use for writing small and self-contained programs, but for real-life and very complex projects the control structures are such as to make the final code almost impossible to comprehend in its entirety.

Methods

In terms of methods, we must distinguish between software methods and application methods. Software methods define the rules on "how to develop a program." Application methods are required for implementing the application.

Let's begin with the more obvious topic — the application methods. They have very little in common with software. For instance, programming a Math library we have to use mathematical formulae — so, it is obvious that we have to know about these mathematical methods instead of trying to reinvent the wheel. In addition, programming any calculation we have to know the idiosyncrasies of discrete mathematics — a belief that common sense may replace the knowledge contained in books written, e.g., by Knuth[4-6] or Graham et al.[7] is just stupidity (there are many books today that may present those topics even better than the cited classic books, but we honor the books that we have on our shelves). When programming a software package for a bank we must have knowledge about the financial world and apply existing methods or processes that are already in use in that environment. When programming industrial control systems, we have to know about the industrial hardware, various physical laws, and control engineering principles. Those requirements are so evident that often companies prefer to employ persons who specialize in the application domain instead of brilliant but innocent programmers for development of their application software.

Software methods should help a programmer to design the architecture of the program, to define the classes or data structures used, and to express the application requirements in terms useful in the coding process. Unfortunately, there is no common understanding or agreement considering software methods. If there were, programmers would be given some education in them, and a person without any knowledge of software methods would be forbidden to code. The reality today is that people who have learned the syntax of a given programming language can call themselves programmers. In reality it is not so easy to solve the split: is it better to take a specialist in the application domain and let the specialist learn programming by doing or to take a true programmer and try to teach the programmer the application problem while programming? The obvious answer — to create a team consisting of both application specialists and true programmers — is in many circumstances difficult to achieve.

The difficulties in defining the skills required of a programmer lead to an extremely large range of programmer efficiency and also of quality of the produced code. It is obvious that complex software developed by persons with a rather vague understanding of software methods is likely to be less reliable than software founded on some established fundamental principles. It is also obvious that developing software without such a foundation often results in projects that are unmanageable and not so rarely end in catastrophe.

Basic Knowledge

What methods does "software engineering" offer? Independently of the programming language used, there is basic knowledge that can be found in any textbook on informatics. Normally, the textbook uses a specific existing programming language for teaching the basics of programming. Of course, we get some mixture of a concrete syntax with general rules, but probably it is better to take this way instead of using some hypothetical language to demonstrate some general concept without having a chance to play with it. Programming cannot be taught as a pure theoretical discipline; programming must include a "learning by doing" component. Dijkstra's thoughts[8,9] about how to solve programming problems without actually programming them is a very nice and idealistic idea, but it will not work with human beings as we are. It makes sense only in relation to (mathematical?) algorithmic problems. Several years ago, Pascal was a very good language for teaching purposes. As structured design and analysis lost its dominant position in programming, C++ replaced Pascal even though something like Smalltalk would be better for teaching purposes. Of course, other languages are also used for teaching the basics of informatics, although using for this purpose, e.g., C or BASIC should be interpreted rather as a misunderstanding.

Informatics courses taught students syntax definition, variables, data structures (records, arrays, lists, classes, pointers, trees), functions, splitting a program into a set of functions, recursive and iterative functions, algorithms (search, sort, hashing), parallel processes, threads, and so on — to list some examples taken from a rather pragmatic book. If we take a more theoretical book we can cite such topics as finite state automata, regular expressions, sets, several sorts of automata (push-down, Turing machines, deterministic, nondeterministic), and so on. There is enough material for teaching quite a few useful concepts, even though some of them may not be very useful in practice. If we realize that there are many persons who program without ever having taken informatics courses, we may express doubts whether education really helps the software industry.

Specifying or Not?

In addition to the basic knowledge, there is another very interesting and important aspect of software methods, which theoretically should be the most important topic of software engineering, namely, the software specification level. The way people develop software oscillates between two extreme positions:

- First specify the programming task and then code it
- Start coding and learn by doing to reach what we really want.

Those two approaches are of course totally incompatible. As those two approaches have survived the entire history of software from its beginning to the present time, both of them must have some rights to be applied.

The specification route seems to be at least theoretically the better way. Comparing software development with other scientific or technical activities, we may argue that basically any construction or production is preceded by a project study. For instance, before we start building a house, an architect prepares the plan of the house and a building engineer does some calculations. But maybe this example is too far away from software specifics, and in any case we have no idea about house building. So, let's take a topic that we know better and compare the development of computer hardware and software. Those are two completely different worlds although they refer to the same product; they are two faces of the same product. Manufacturing of electronic hardware is a well-organized and planned process. Hardware comes into being on paper (effectively in CAD programs): the electrical schemes, the layouts, simulation, checking — several steps that produce the entire project of the hardware. Only when the hardware is entirely designed (in other words, completely specified) may the manufacturing start. Considering that manufacturing is a highly automated process, the only way to do it successfully is to have a complete project for the production line. Contrary to that, software development does not require any design or specification phase — we may start manufacturing (coding) while having a very vague understanding of the application. If we may do something it does not mean that we should do it but the problem lies in the possibility:

We may do it that way, hence we do it that way.

The protagonists of "learning by doing" software development argue that customers do not know exactly what they want, or at least they cannot express it in a complete fashion. But the reason for the customers' attitude to their requirement is exactly the same as for the programmers: the customers know that they are not forced to supply a precise specification so they economize. We are here again in the situation: we may do it, hence we do it. We could formulate a law that applies to both the software development team and the customer:

We choose always the way of apparent convenience.

Let us have a look at the most prominent representatives of those two different software development methods: first CASE tools, especially UML, and then Agile methods.

CASE Tools

CASE (Computer-Aided Software Engineering) tools are nearly as old as the software industry. From the beginning there was an awareness of the need to support programming by automated tools. Hence, under this very broad topic, thousands of methods and tools have been developed for any imaginable software activity. The list of supporting tools is very large and contains program simulation and verification, application analysis and modeling, software architecture modeling, machine code analysis, code review, data and database modeling, design documentation, specification and executable specification, executable model, object-oriented analysis and design, project management, reverse engineering, structured analysis and design, to cite some rather randomly chosen examples.

We exclude from that topic the programs needed to generate the machine code, such as assemblers, compilers, linkers, loaders. We also exclude editors, debuggers, or code analyzers, which we consider as development tools.

Usually, CASE tools have powerful graphic editors, which makes it easy to draw nice diagrams. The formal methods that we need for a design are missing. The result is then a document that contains diagrams and text; in other words, the requirements in a better presentation. Many programmers struggle against that. They agree that on the way they come to better understand the requirements, but they consider the time investment too high. Partly we can understand that attitude, although in many cases it is just an excuse. The true reason for refusing the specification phase is often that programmers like coding. Specification work is obviously not fun. The other reasons for the negative attitude against CASE tools is that the awaited advantages of the (formal) specification, like verification against logical errors and consistency of the entire specification, cannot be fulfilled.

To improve the usefulness of the specification, the CASE tools try to generate code. This direction seems to be the greatest misunderstanding: code generation makes sense only if it is complete, but then we do not speak any longer about a specification; rather, we will have a new programming language. More comments about this issue can be found below in relation to UML.

UML

For the purpose of the book we limit our discussion to CASE tools that started in parallel to the introduction of object-oriented design, the governing software paradigm. For several years, several groups have tried to convince programmers about their notations but failed. Eventually, they agreed to a common notation under the name UML, which stands for Unified Modeling Language. Unfortunately, instead of creating a synthesis of their many years experience, they merged several not-quite-compatible concepts and notations, producing a monster whose usefulness should be at least critically discussed. Especially, we can note that UML has been in use for some years, and there is no hard evidence that the use of UML makes an object-oriented design successful.

But what kind of method is behind the notation? This is probably the weakest point of UML. It is just a notation of the object-oriented design but it does not imply any method. Is a definition of some symbols for graphical representation a method? Some parts of UML can be described as a method; a good example is Harel's Statecharts[10] method for presenting sequential processes. But as a whole, UML presents a rather ill-defined concept.

UML has several weak points that result from its overloading with elements that overlap and confuse the user. To be more specific, let us take presentations of sequential activities. There are three different diagram types to show more or less the same aspect: state diagram (another name for Statecharts), sequence diagram (a kind of time diagram), and activity diagram (actually a flowchart). The point here is that although they relate to the same topic, they are not equivalent and not convertible to other presentations. Hence, we have to draw all those diagrams by hand if we need a complete presentation of the "dynamic" part of the specification. So, do the diagrams present the same thing or not? Rather not: they present different aspects of software behavior. Wouldn't it be better to have one diagram for sequential activities? This is an effect of producing a universal notation by merging several notations.

The other topic that should be addressed is the alleged universality of the notation. Theoretically, an object-oriented design method should be independent of the application domain. To some extent it is true but it has one drawback. Business applications that are at the origin of the notation do not know control problems, as understood, e.g., in industrial control or telecommunications. Therefore, the notation is relatively good considering data presentation and weak considering control.

As we do not want to make any serious analysis of UML we limit ourselves to citation of some excerpts from an opinion that I wrote some time ago about one of the UML tools:

> The principal problem of any CASE-tool is that they are not a full programming language. Hence, they allow realizing only a part of the development work (specification, modeling). This means that the result of a specification which is done with a CASE-tool must be completed with other means (programming languages). There are possibilities (Reverse Engineering) to synchronize the further development with the original incomplete specification but in practice it does not function and is used only to demonstrate the tool features. This does not work not only due to shortcomings of the tools that are not able to do it reliable and fully. If a programmer takes over the specification results and starts to program it becomes very difficult for him to "waste" time on things which are negligible for him. This typical psychological effect is well shown by notoriously bad software documentation.
>
> The usage of the tool should lead to better software documentation. To analyze whether it is true, let us take an ideal case of software development where the project has been successfully terminated and the modeling documents have been actualized with help of Reverse Engineering. Let us ask two questions.
>
> The first question is what kind of documents will be read to understand a program. UML documents are understood by persons who know the modeling language, just like a C++ program can be read, to some extent, by a C++ programmer. We should not be deceived by statements that graphical presentations are more expressive than a code, especially if stated

by people who know neither UML notation nor C++. If we see some Japanese writing or Egyptian hieroglyphs we are impressed by their graphical beauty but we do not understand them. Considering UML and similar notations: how can a person understand a UML diagram without knowing the true meaning of, e.g., 10 different arrow types that are used on the drawing? Only informal drawings where we use not precisely defined squares, circles, lines and arrows are "understandable." A precise diagram where each element has an exact meaning can be as difficult to read as any other code and can be read only by qualified persons.

The second question is, what kind of documents are we going to use in software debugging or changes? The answer is rather obvious: we use C++ code for these activities.

What do we really need modeling documents for? They are useful in the beginning of software development to discuss ideas and concepts. They make it easy to introduce new programmers into a project.

UML and programming languages are not the same tools. When the software model is ready, the UML tool generates for a C++ implementation h- and cpp-files; the cpp-files are only dummies with tool-specific headings. As it is a general tool there is no one-to-one correspondence between the modeling language and the C++ structures. This means that the h-files generated by tools are not good (e.g., sometimes they do not know references). To allow for reverse engineering the h-files are overloaded with tool specific comments which make them unreadable (we know the problem for any wizard-generated files).

Since software is developed with two different tools (modeling and programming) it requires two different development environments. If the two tasks (modeling and programming) are performed by two different persons they will be effective as each of them works in a well known environment. This solution requires good cooperation between these two persons because modeling and programming are not completely independent tasks and require interaction and fine tuning. If both tasks are performed by the same person the load and quality of the person must be higher as this person must use two different tools.

Are there any alternatives to modeling with UML or similar tools? Should we continue in the old manner even though we know well the shortcomings of the software development process? I think if we invest time into software specification and documentation instead into modeling with UML we get similar results without creating the impression that we get a new quality using UML tool. If we invest time in software documentation we produce software which is easy to maintain and change. Do we have any other aim?

CASE-tools will bring a new quality into software development if they are truly programming languages which allow the application to be completely specified and generate an executable code. To realize it requires also creating a debug environment. All this is very complicated and therefore we have still to wait a while before we get useful CASE-tools for software development.

That comment was written some years ago. The tools are getting better but the improvements do not change the general situation. The prospects shown by Edwards[11] do not seem very promising. To clear any doubts after these rather critical words: we are convinced that software cannot be developed without well-defined requirements, specification, analysis, and design. Translating of the specification into code should be done at the moment when we are pretty sure that we really know the application and nearly all doubts and details are cleared; ideally the code should be produced automatically. But we do not see in UML the realization of such defined goals.

Agile Methods

Let us now have a look at the opposite approach, categorized as Agile methods. In its essence, an Agile method is based on a set of well-meant recommendations related to coding or programmers' cooperation in the belief that as a side effect it renders specification and design largely unnecessary. In fact, reading the Manifesto of the Agile movement we agree with much of its content, but we know also that it will not work. To be accepted and successful any method or tool must be made for people (the communist ideal was also fine but it had one weak point — it had been formulated without taking into consideration the nature of human beings). To assume that we have a team of perfect programmers willing and working smoothly to get software ready in time is a very nice proclamation, which will be signed by any manager but is too naive and has little to do with real life. Software development cannot be based solely on good will and cooperation of engaged persons. It also requires methods, planning, schedules, and organization. The basic Agile recommendations are very vague and in fact of such general nature that they do not fit well into a technical book. They could have been produced by any manager who wants to motivate his or her employees. To be more specific let us cite the most important rules of eXtreme Programming (see Table 1.1), which is a well-known implementation of the Agile idea (in the right column we put our comments).

Perhaps some of our comments are not quite fair, but should we really accept and even discuss any proposal which is so foreign to real life? There is an explanation for this kind of "method." The explanation stems from the definition of software that we discuss later. We mention here only that there are no known large software projects that successfully use the eXtreme Programming idea — the explanation is linked to the size of the developed software. We found the following characterization of Agile methods by Whittaker[12]:

> If taken "to the extreme," agile development is a completely unstructured, chaotic process that employs unrepeatable processes and bypasses much of the testing and design phases.

Whether these kinds of methods can really contribute to software development can be at least questioned.

We have presented here the two extreme ways of developing software and expressed our critical opinion about them. Of course, the reality is never as extreme as the used methods suggest. Programmers must deliver software. They are bound to deadlines and requirements that should be fulfilled. To develop software in time and in the budget frame they choose what they consider the best solution for them. Good management may help them but eventually the programmers take the full responsibility for success or failure. And good management will never try to enforce any method that is not accepted by programmers. Bad management does it sometimes and programmers behave in such a case as if they use the method, producing, e.g., some useless documents that fulfill the formal requirements of the enforced methodology. But the true programming work is done in parallel to that fictive work.

Behavior Modeling

Software always articulates itself by behavior. There is software whose behavior is simple but there is also software whose essence is the behavior. For instance, applications that control something are defined by the

Table 1.1   eXtreme Programming Rule

eXtreme Programming Rule	Our Comment
Two programmers work on each task using a common keyboard and a monitor: one writes the code, the other one thinks and helps.	How many programmers exist who will voluntary accept this kind of work?
Any ready software component is immediately integrated into the built software.	Integration of a major software product is a complicated task normally requiring many hours, sometimes days. Therefore, it cannot and must not be done on a daily basis.
Test units will be written for any piece of software before the actual pieces of software are coded.	How can one write a test of software pieces without having a perfect specification of each of those pieces?
The client determines the goals of software test and does it in person.	One of the reasons for use of eXtreme Programming is the fact that clients as a rule do not know exactly what they want. Hence, how can they test the intermediate results and accept them without knowing what they really intend to have? How many clients are known who want to take part in the development process?
The software architecture and software documentation are continuously improved.	The main reasons for confusion in software development are changes of the existing parts. Therefore, all changes in a software project must be well controlled and coordinated. As a rule: the fewer changes, the better.
The working time is limited to 40 hours.	It might be a good law for a labor union.

behavior description of the controlled process. Therefore, methods allowing modeling of behavior are used in software specification and design. There are several methods used for that purpose; the best-known are finite state machines, Petri nets, and Statecharts. All those methods use the concept of a state* to describe a present situation, which is defined not only by the current inputs but also by the history of input changes.

Behavior modeling methods are to a large extent underestimated in software, as they are considered too theoretical and inefficient, especially in software design. Misunderstandings surrounding that topic result in poor design where the history of input events is "stored" by flags, markers, and similar variables, which cannot be well managed and are responsible for most software disasters.

Development Tools

Development tools are means used in programming — writing the source code and creating the executable machine code. In previous decades widely different development worlds have been established, with some-times contrary design philosophies. Any discussion about development tools must also contain the different ideas about software ownership and marketing. It cannot be limited to a purely technical review of editors and compilers. Otherwise, it would be difficult to understand why in the 21st century we have programmers who still write code using the editor vi coexisting with programmers using a powerful IDE (standing for Integrated Development Environment).

To effectively produce reliable code requires several programs. We may leave the use and coordination of these programs to a programmer or we may create a development environment (IDE) where the chain of activities is automated. The activities contain, among others:

- Writing a source code
- Creating GUI
- Compiling
- Debugging: step, trace, spy
- Optimizing code
- Performance analysis
- Linking
- Producing loadable modules
- Producing installation package
- Managing source code (versions)
- Generating documentation

---

* A *place* in a Petri net is not like a *state* of an fsm but a Petri net may be transformed to an equivalent system of fsms.

Programs that perform the above tasks must cooperate with specification tools if they are used (e.g., UML) and must be supported by Help and Knowledge Databases (local and Internet). The list can be longer depending on the manufacturer.

Looking at the list of tasks required for producing a program we understand that a good IDE can contribute to programmers' effectiveness. The desktop software development platform is dominated by a few big companies, on top Microsoft Visual Studio (for several languages like C, C++, C#, J#, Visual Basic, but only for Windows) followed by Borland's Delphi (only Pascal but for Windows and Linux). There are also some platform-independent IDEs, e.g., Eclipse.

Creation of a useful and reliable set of programs required by an IDE demands huge resources, which explains why only big teams (companies) can manage it. The development of an IDE is a process that takes several years, but these days IDEs are very powerful platforms, which fulfill nearly all programmers' wishes — they are really good, contributing a lot to software development.

The other extreme is to control the entire development process by the programmer. For instance, a programmer who works in the Open Source environment is confronted with less-user-friendly tools. The necessity of using relatively primitive editors, debuggers, and working on the command line is considered by some over-enthusiastic GNU fans as the proof of being a "true" programmer. To some extent it is true that programming in the GNU environment requires more engagement for project management, writing Makefiles, doing a lot by hand or in general being deeply involved in all details of the software creation process. It could be also considered as a waste of time. But the situation improves and such emerging tools as, e.g., Kdevelop or Eclipse, are becoming the IDE for GNU. This is very much required to relieve programmers from dealing with all the development details, which delays the programming work.

Recommended Reading

1. Baldwin, G., "Toward an Assembly Language Standard," *IEEE Micro* (August 1984): 82–85.
2. Nicoud, J. D., Wagner F., *Major Microprocessors: A Unified Approach Using CALM*. Amsterdam: North-Holland, 1987.
3. Dijkstra, E. W., "GOTO considered harmful," letter, 1966.
4. Knuth, D. E., *The Art of Computer Programming*, Vol. 1, *Fundamental Algorithms*. Reading, MA: Addison-Wesley, 1968.
5. Knuth, D. E., *The Art of Computer Programming*, Vol. 2, *Seminumerical Algorithms*. Reading, MA: Addison-Wesley, 1969.

6. Knuth, D. E., *The Art of Computer Programming*, Vol. 3, *Sorting and Searching*. Reading, MA: Addison-Wesley, 1973.

7. Graham, R. L., Knuth, D. E., Patashnik, O., *Concrete Mathematics: A Foundation for Computer Science*. Reading, MA: Addison-Wesley, 1988.

8. Dijkstra, E. W., "On the cruelty of really teaching computer science," *Communication of the ACM* (November 1989).

9. Dijkstra, E. W., *A Discipline of Programming*. Englewood Cliffs, NJ: Prentice-Hall, 1976.

10. Harel, D., "Statecharts: A Visual Formalism for Complex Systems," *Science of Computer Programming* 8 (June 1987): 231–274.

11. Edwards, C., "Modeling standard gets ready for second round," *IEE Electronics Systems and Software* (October/November 2003): 36–39.

12. Whittaker, J. A., Voas, J. M., "50 years of software: Key principle for quality," *IT Pro* (November/December 2002): 28–35.

Chapter 2

The Price of Weakness

Software Development Costs

Software development costs are high and are continuously growing. The costs increase over-proportionally to increasing requirements. Unreasonably high costs are accepted as a native feature of software; the reasons are manifold:

- The development of programming languages stagnates.
- There are no convincing methods that truly accelerate software development.
- The software part is getting "harder."
- The alleged freedom in software development leads often to an unnecessary proprietary solution.
- The management of software development is very often completely uncoupled from the real programmers' world.
- Trial and error methods replace designing.
- Lack of exact specifications makes software maintenance and enhancement very time-consuming.
- A mixture of several programming languages and (proprietary) style standards makes it difficult to introduce a newcomer into the project.
- Too many surprises due to lack of clear requirements and specification.

The present variety of programming languages used reflects first of all the situation that there are no universal programming languages: different application domains require specific languages. But do we really have hundreds of application domains?

Before we look more closely at the reasons for constantly increasing software development costs we should like to define what we really mean when speaking about software. Many misunderstandings start with the lack of agreement: where the topic "software development" actually begins. To expose better the problem we define three groups of software: hobby, small, and large projects.

Once I read a nice book written by a famous physicist. A group of physicists had tried to understand some physical phenomena and they needed software to calculate some formulas. One of their friends (also a physicist) had been considered a software specialist, but unfortunately he did not like to work on weekends. Eventually, they convinced him to come on a Saturday afternoon and he managed to develop in a few hours a fantastic software package which helped them very much in their work. The book is full of ironic comments and presentation but just this fragment is serious — the author was convinced that he had experienced here how software is written by talented men who do it just "like this." The story demonstrates the ideas non-programmers have about software — to write a useful program in a few hours.

Programming as a Hobby

In any textbook we find some sample programs. Students write programs as seminar or diploma work. Some people have as a hobby writing programs to do some useful job (see, e.g., freeware and most of shareware software). Persons making some measurements create script programs automating the measurement processes. Several scripts or macros, which require thousands of willing persons considering their variety, belong to that category — students do those tasks willingly to earn their living. All of those activities can be characterized as "any programming language and development tool used will do."

In all those software projects the responsibility is negligible and they are not determined by any marketing considerations. The main goals of such programming are of personal character, e.g., the "joy of coding." That sort of programming is done by one person. There is no necessity or pressure for any organization, methodology, documentation, testing procedures, etc. The result reflects the abilities and personality of the persons who have done the job — often very well. It is not software development in its true meaning — it is just coding (we should not understand here "coding" in a pejorative sense: to produce a good code requires skill).

Small Software Projects

Today, nearly all products (kitchen accessories, toys, DVD recorders, or industrial equipment such as motor drives, measurement instruments, etc.: the list is endless) require and contain software. Those software projects have a manageable size; to be more specific, they can be realized by one or two persons during a period of a few months. The basic difference between those small software projects and hobby projects is that marketing is involved; i.e., the product must be defined by requirements, programming should be done in a certain time and budget frame, the version and maintenance management cannot be ignored. If such a project is done by one person it is still a "one man show," being very effective but of course with full dependency on one person. The addition of the second programmer changes rapidly the situation — it is just a completely different environment. Human factors play a very important role — it would be an illusion to believe that two programmers mean double efficiency (see also Figure 2.1). It may be true in some rare cases but as a rule management has to foresee overhead costs resulting from work organization and communication between two persons. It is a very serious problem, which means that companies try as much as possible to keep the "one man shows" as long as possible. It is more effective to have two different small projects done separately by two programmers instead of arranging that both programmers would work on both projects. The bad side of such practice is that there are different programming styles in the same company, which makes maintenance even more expensive.

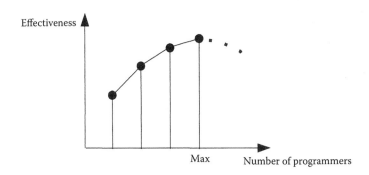

Figure 2.1 Effectiveness of programmers' group.

Large Software Projects

The next step means that the software projects are large, involving tens and sometimes hundreds of persons. We write here intentionally "persons" and not "programmers": large software projects require not only programmers but also people designing the graphical user interface (GUI), writing help manuals, testing, building the application, last but not least managers who organize the project because the work of all the participants requires coordination. In those projects all components of software development play a role: a specification method, a programming language, and a development environment. The organization of the project can be a very complex task, and often this is just the factor that determines failure or success.

Programming done in large teams exceeds by far any experience of persons who were never involved in such a software adventure. The continued conflicts between requirements or specification and code, the synchronization of several parts written in parallel, the management's (justified) desire to keep control over the process (plans, schedules, milestones, code reviews, and never-ending meetings); all this creates an atmosphere of continuously growing tensions which unloads itself around milestones and eventually explodes by the deadline where the truth (is it done or not?) cannot be denied anymore. Of course, we may say that those symptoms are typical of any large project (buildings, factories, bridges, etc.) but software development has two characteristic features that are not known in other industries.

First, progress is actually invisible. While building a bridge we may assess every day the progress, even with the naked eye. While developing software, even closely involved persons have difficulty making any reliable estimation or prediction. For the outside world, including management, the progress is hidden. Numbers of written code lines, passed tests, done code reviews are elements of the fictive world, which is there to convince management of the progress. For instance, spending a day adding comments to the code greatly improves certain software statistics. Hence, it is not rare that the fiasco of the project is only disclosed at the last moment when it is too late to react. It explains also why tools that fake the reality are so popular. For instance, a "good" tool or project leader should first produce the GUI. This makes a very good impression as it is at least visible. That there is nothing substantial behind that façade is not important — obviously to live with an illusion is easier than to try to understand the truth.

The second software specific is that software development never turns into software manufacturing. Most software has a prototype character. Developing a car terminates with a product that is then sold "as is" in many thousands, sometimes millions of examples. The car type goes into production when very intensive tests prove that it is reliable, without such failures that would make it impossible to function. In most cases it works: recall actions are rare and can cost car manufacturers a fortune. Software is sold if it reaches certain acceptance tests, even though we know that it still contains errors that make it impossible to function properly in certain situations. But selling thousands of CDs with a program is not comparable with selling cars. The CDs will never be recalled even though they contain several bugs. With continuous patches and updates companies try to improve the software, but several examples show that after a while the number of errors stays constant — obviously removing errors produces the same amount of new errors, in other words developers are not able to improve the software anymore. The patches and updates reach only some of the customers — look at the state of the operating system in most PCs.

In relation to points made in the preceding paragraph it is fascinating to see that car manufacturers are beginning to have very serious problems with the software they are installing. They are becoming responsible for products that have a very significant software element, but these companies have no suitable "culture," i.e., no appreciable background in developing software. As the difficulties of integrating poorly specified components into the overall system increase, these companies are tempted to take panic measures, such as recruitment of another 300 programmers, mostly inexperienced. Unless significantly improved design methods are adopted, we shall soon see some disasters dwarfing what we have seen in the past few years.

We know of software that has been officially "improved" for many years, but the only effects of those improvements are additional functions that nobody needs: the essence of its problems stays untouched. As an example we may mention Microsoft Word, which has basic problems in handling large and complex documents — after a certain size or number of graphics the document becomes unmanageable. Probably the program lost completely its initial concept, reached its limits, and cannot be improved. We may risk a forecast that Word will forever stay a program to write letters and not complex documents — the only way to cure it would be to write it from scratch defining a new concept adequate to the goal and using appropriate methods. In the same state are several popular user programs (money management, tax declaration, etc.); the situation there is worse because of the limited resources available to the programming teams.

The application programs we use in everyday life, e.g., text editors, spreadsheets, or Internet browsers, are relatively well tested: we millions of users serve as guinea pigs, supplying the software companies with bug information. But such programs are rarely without faults, even after the sixth or seventh major revision. There are also programs that are written just for a single customer or very few customers. Those programs are just very poorly tested prototypes because they never see a sufficient variety of test situations — compare them with the above-mentioned desktop applications. Interestingly enough, customers seem not to realize the situation and assume that it is possible to write reliable software without corresponding testing effort and expect reliability exceeding that of the desktop application they use in everyday life.

A quite large percentage of software projects end with a catastrophe: either the development team must admit that it cannot achieve the goals or the expenses and delays are beyond any acceptable and foreseen limits and the investors stop the financing and terminate the project. If we put aside at this moment human inabilities or criminal activities and take into consideration only the technical components, such software disasters result from underestimation of the problem (naivety) or because the tools used have reached their limits. Unfortunately, the human resources — the size of programming teams — cannot be increased endlessly (see Figure 2.1). We cannot increase the size of software groups beyond certain limits. Somewhere between five and ten is the absolute maximum size of a programmers' group. Any increase beyond that number produces more harm than advantages. We speak about a group working together on the same closely coupled software modules. Fortunately, very large software problems can and must be partitioned into isolated software tasks, which communicate via clean interfaces so that they do not influence and disturb each other. The barriers that limit the possibilities of increasing the manpower of software groups have been recognized very early and were well expressed already in the sixties by the Brook's law[1]:

Adding manpower to a late software project makes it later.

Taking the most important software features: functionality, user friendliness, reliability, and maintainability, the requirements are increasing continuously. We can fulfill the requirements either by using better, more effective development tools or by "brute force" (more manpower). The very important factor in the development environment — the programming languages stagnate, which is demonstrated by very curious phenomena, like hardness of the "software" concept and ease of creating new macro languages.

Hardness of Software

Presenting the short history of programming languages we mentioned that the new languages had never truly replaced the old ones. Obviously, the cost of a change is considered higher than the expected revenues. There are, of course, reasons not to do it because there is no universal language. For instance, it might not make sense to use C++ for programming of some tiny embedded system based on a PIC micro-controller. But even if we take application domains that could be covered by certain language types, there are no obvious technical reasons to change the programming language because the expected revenues are rather illusory. So, a change, e.g., from Pascal to C or from C++ to Java will be done in a company due to some personal preferences and not as a result of any evaluation.

The power of the bounds to a certain programming language can be well demonstrated on PLC. The PLC programming concept stems from a hardware realization of control circuits in the fifties, which were done using electromechanical relays. Those control circuits have been presented as ladder diagrams where, e.g., two relay contacts connected in series realized the AND Boolean function, two relay contacts connected in parallel realized the OR Boolean function, and a flip-flop could have been realized as a combination of series and parallel connections — all shown in Figure 2.2. Those first realizations of digital circuits have been long superseded in hardware design. Probably, today students do not even hear about those prehistoric solutions of control circuits. Interestingly, it stays as a basis of the PLC programming environment — there are, of course, several presentations built on it but the concept has not been changed. It is difficult to find any reasonable technical justification why we should still be bound to some concept which made sense 50 years ago but has completely lost its validity now. That example is an interesting element showing the difference between hardware and software development — the alleged "soft" programming methods once introduced are very difficult to change; in contrast, hardware adjusts and uses possibilities offered by technological progress.

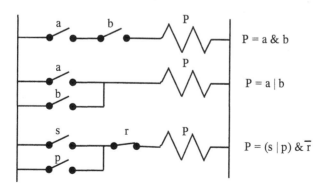

Figure 2.2   Relay representation of Boolean expressions: AND, OR, R-S flip-flop.

The other examples that illustrate the hardness of software are old programs, especially libraries used for decades. On the one hand, they are very reliable components, having been tested for a long time. On the other hand, they cannot be improved even if their errors are known — as a side effect such a change could ruin too much software based on it. Often they do not fit into the environment of more modern languages. For instance, the use of the C standard library should be avoided if not forbidden when programming in C++ (see, for example, the string-handling problem mentioned elsewhere), but that is sometimes impossible. Hence, we have to live with inconveniences that as a whole contribute to software problems.

Ease of Creating New Macro Languages

The other factor that contributes to the inefficiency of programming is the ease of producing one's own macro or script language. These are created either by ambitious programmers or they just arise naturally. We illustrate it by a case from our experience. A StateWORKS tool that is discussed in Part III of this book needed a command language for automating tests and allow them to be repeated without manual work. As the communication with the StateWORKS runtime system is a typical server–client link over TCP/IP the base of the command language has been easy to define: the messages of the TCP/IP connection. Defining simple syntax rules has led to a definition of several commands that are nothing other than the TCP/IP messages in an appropriate cover, like: connect, disconnect, get, set, etc. The next "obvious" step would be to introduce a few typical conditional (if-then-else) and loop (for, while) statements, and a new macro language would be born. This scheme has been repeated all the time; why not do it again producing another macro language? Once starting in this direction we could eventually end with a quite sophisticated language containing all imaginable fireworks and we could use nice marketing arguments praising the alleged superiority of our command language over competitors' solutions. We have not done it for reasons of principle as we consider this to be a totally unproductive activity. Our so-called command language is to be applied for creating command files from log files that are produced during testing. Of course, we can write the command files by hand in any text editor and if we want to go a step farther we have enough programming languages with nice conditional and loop statements to handle our strings. So, being sincere there is no reason to expand our command language beyond the few statements representing the TCP/IP messages. To be frank, we also must admit that as programmers we would like to do it — coding is just fascinating.

A similar situation exists in most proprietary script and macro languages in use today. The reasons for their creation are vague and lost in the past. Is it really disturbing? Does the pressure of the marketing division to produce all the time proprietary solutions, differing from those of the competitors, justify these kinds of achievements? The answer is to some degree yes; eventually our products must be sold. But we should not forget the price we pay for those activities, a price that, just in software, is enormous. Any proprietary solution that isolates us from the world means that any new employee has to be taught our tools, and our language requires our testing and debugging tools, our maintenance, and documentation. If the technology changes we may be forced to redesign our programming language. Because of the exorbitant cost of such proprietary solutions large companies try to sell them eventually as a product. Very few of them succeed.

Do We Need So Many Programming Languages?

The answer would be the same as for the question: Do we really need so many types of cars or any other gadgets that we see in our shops? Starting from some idealistic but unrealistic assumptions we could probably prove that we need neither so many car types nor so many programming language types. We are living in a capitalist environment that encourages both innovation and market dominance, even though these seem in contradiction. This obvious and prosaic answer does not mean that we should not comment — and still less accept the situation. We do not want to discuss here the commercial wars between large companies, which generate products from a marketing point of view — at the end they lose if they exaggerate their marketing activities to the detriment of technical progress. We have experienced enough examples of this sort in the past.

The serious question is: Do we see any progress in programming languages? We have serious doubts about that or, frankly speaking, we are convinced that there has been no progress in that field since the invention of the object-oriented paradigm. We are confronted of course every few years with a new language, but the principles do not change. A programmer must still convert very complex requirements into a working application using primitive instruction sets. The resulting code is extremely hard to read, let alone analyze to study its behavior in all situations. The development environments of programming languages offer large and sophisticated libraries to support programmers' work: MFC libraries in Microsoft Visual Studio or STL library are good examples of them. Without such libraries, today's programming tasks would not be realizable. These additional means do not change the rather pessimistic picture, which is characterized by a lack of genuine progress. Fascination with developing new programming languages results in some occasional improvements but effectively those do not change the situation.

We presented the way the programming languages evolved. Since object-oriented languages, e.g., C++, are the programming languages of the present time, it does not mean that the whole software world programs in those languages. Nearly the entire palette of languages ever developed is still in use: assemblers, BASIC, Pascal, COBOL, PLC, and so on. It seems that a language, once introduced, only dies with the programmers or the company — the change to a new language is very difficult and seldom done voluntarily.

The Specifics of Programming Languages

Modern programming languages spread the impression that it is so easy to write a program. We write the first line of code, followed by the second line — we compile it and we see that it is correct. We are sure that we can continue endlessly in this style although we learn from our own and other programmers' experience that with each line of code the probability of error increases dramatically. Let us take a randomly chosen piece of code:

```
if ( m_Client.Request(m_Name, m_Attribute, m_Value) )
{
    if ( !m_Value.empty() )
    {
        m_asObjects[i] = m_Value;
    }
}
```

If we really have the time and are in a mood to do it we should check:

- Are the pointers valid?
- Do parameters of the Request function have valid values or does the function double-check it?
- Is the index *i* valid?
- Is the returned value *m_Value* valid?
- What should be done if the outer *if* statement fails?
- What should be done if the inner *if* statement fails?

Those questions are not answered in the code. In most cases it works because the context assures that the values, pointers, etc. are valid and that ignoring conditions that are not fulfilled is harmless, in most cases.

But the entire code of a program is constructed in principle in such a way — it contains thousands, sometimes millions of code lines for which correctness is based on the principle that the context is trusted to be correct. From time to time it is not correct and the software crashes.

The other specific of modern programming languages is that programs are mixtures of very primitive and very sophisticated statements. For instance, one of the "favorite" programmers' occupations in C has been (and continues to be in C++) processing of strings. The difficulty is inherently bound to the different representations of strings as a pointer to character, array of characters, and generic string. The generic string also may have different definitions, e.g., defined as STL or MFC type. If all those definitions are used in the same software (which is quite normal considering the dependencies on established libraries) the code is full of explicit and implicit conversions and type castings. The consequence is that on the one hand we use, e.g., very powerful and elegant STL methods and algorithms, but before we can call them we have to "prepare" a character or string to be usable for them. From that point of view the code has sometimes a rather sad form and what is more important — the cost of developing such a code is very high, the proper solution being found often after several trials and not as a result of careful planning. The reliability of such code is also low as the side effects resulting from those character and string manipulations are sometimes unpredictable and occur when the code is already in use by a customer. Then, last but not least — those "exercises" are repeated all over the world by thousands of programmers every day.

The Specifics of a Software Project

In Figure 2.3 we show a dependency that well illustrates what is perhaps the basic reason for delays of software projects, their costs, and failures.

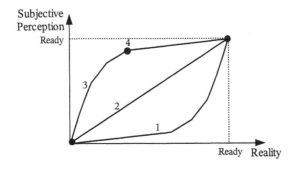

Figure 2.3   Estimation of software project progress.

The essence of the diagram is the finding that the end of the software project is not well defined. We exclude in these considerations the deviations caused by human characters, which cause one person to have a more optimistic and another person a more pessimistic view on the same subject.

Curve 2 shows an idealized course of a project where the subjective perception of the participant corresponds to the real progress.

Curve 1 represents projects where a participant is convinced most of the time that the project is delayed, being at the end surprised that everything ended well. Writing a book is a good example of such an undertaking. We can be overly optimistic only for a first book. We never forget our first experience and we expect that the work will never end. We do not believe in gauging progress by counting the number of written pages as we know that on the next day we will decide to rewrite some of them and this procedure will be repeated several times. We do not forget the never-ending discussions with our friends and co-authors who find new errors or have new concepts. We do not forget the auxiliary activities that always cost us more time than planned: figures, diagrams, tables, appendices, and the fight with the editing tools (which demonstrate to us all the time the weakness of software products). Last but not the least is the satisfaction with the written text — we are rarely really satisfied with the work done and we have the impression that we should rewrite it. And suddenly, when we have generated the index we discover that the work is finished.

Curve 3 shows the typical feeling of a programmer who demonstrates an over-optimistic view in the first phase of the project. The programmer is convinced that the project is nearly ready around point 4. A more optimistic person may even signal to the chief: "My program is in pretty good shape; what else can I do?" The reality is, as a rule, much worse — most of the (unpleasant) work: the details, debugging, testing, documentation, performance, etc. still remains to be done and may cause a feedback with disastrous effects. The difficulty increases because of the not well-defined end of the project; e.g., does maintenance belong to the project? In contrast to a book writer, a programmer never learns from the past. The new project is always the beginning of a new software era: this time it will be done well, according to the time schedules and in the financial frame. As programmers we tend to underestimate the programming effort required and overestimate our resources and skill.

Especially pronounced is the problem of software outsourcing. The supplying company is in the convenient situation that officially its obligation ends if the requirements are fulfilled. As the requirements are never complete and do not cover most of the work that must be done over point 4 the outsourcing company is able to deliver the product relatively soon. The first tests on the customer site disclose the missing requirements and the result often threatens a financial disaster depending on the contract and involving lawyers.

Software Is Expensive

In the beginning of the computer industry, hardware costs were very high and in relation to them software expenses seemed to be moderate. The relation between hardware and software costs has changed all the time to software's disadvantage. Hardware becomes more powerful and cheaper. Software development tools fell behind the increasing requirements, which amplify the trend of increasing software development prices. The ever-growing software development expenses can be compensated for by the growing market. Therefore, mass-market software, such as operating systems and most of the desktop application (text editors, spreadsheets, and several utilities), is cheap as the amount of copies sold covers the expenses. In contrast, the cost of custom tailored software reaches exorbitant levels. Because this process has continued for many years, the whole world gets accustomed to it and accepts it as a software feature — we assume that it cannot be changed.

Maintenance Costs

Software does not wear out as mechanical or electrical devices do. In spite of that nice feature software maintenance is an important and expensive activity. Because maintenance expenses are in many cases the main cost element of software we should take a closer look at it. There are two reasons for maintenance: errors and changes.

Software Errors

Errors are called *bugs* in software. Errors cannot be totally eliminated from programs. Any software contains some "known bugs" and a certain number of unknown errors. The way software is designed and coded is very much dependent on a programmer's skill and it is in fact "hand made." Hence, it must contain errors. Automation of routine programming tasks, automation of test procedures, usage of proven libraries supported by formal design and specification methods may reduce the number of errors. At that moment programmers concentrate mostly on debugging, that is, on searching for errors in code. When software reaches customers, the search for errors continues and is then called maintenance. During the maintenance the customers play a very important role in supplying information about bugs.

There are two types of error. Let us call them: *software bugs* and *application errors*. Software bugs are errors related to the programming language and they are made by programmers while coding (misspelling statements or variables, invalid values, logical errors, deadlocks, loops, wrong constants, wrong addresses, etc.) or by translation programs (compiler, linker, loader: rather rarely today). Those bugs are the dominating errors in some programs such as text editors or translation programs or general programs, which are developed and updated over a period of many years. For those programs the application specification is known for many years and software companies have problems to justify pseudo-improvements and extensions that may bring money — elimination of bugs cannot be sold as an improvement.

Application errors are of a different nature. They result from erroneous or contradictory requirements, misinterpretation of specifications, or just application logical errors during coding. Application errors are a huge problem in programs that are written for a specific application, often for a single customer.

In case of software misbehavior it is often impossible to identify the error type: software bug or application error. Software crashes indicate clearly a software bug. A wrong robot movement may be caused by any sort of error. Software bugs can be found only in code. Application errors can be found in code or in a specification.

Software Changes

Maintenance also means changes. Even if software development is based on well-defined requirements, it is just impossible to produce software that fulfils 100% of a customer's expectations. First, requirements are verbal documents whose content is interpreted — programmers sometimes misunderstand requirements. Second, it is only by using the software that customers learn what was wrong in their specification and demand changes.

Maintenance is a difficult process as it means changes to running software. Even if the changes themselves are not so complex the replacement of the software at a customer's premises may be a nightmare, especially if it is left to the customers. In most cases, software must be exchanged by a specialist.

Software can be changed and adapted to new requirements only to some extent; its limits are defined by its design and by design corruptions done during coding. While coding we always meet situations that were not foreseen by the original software architect. We can either change the design or bypass the rules and solve the problem by using some dirty tricks. We tend to do that "just once" the first time, being convinced that it is only an exception that will not spoil the nice architecture — and anyway we are in a hurry at that moment and maybe we will improve it some time later. We do a few such dirty tricks several times, so corrupting the software. Eventually the software does not have any coherent architecture and documents describing its design will have completely lost their value. I once did a search for the word "friend" in some software whose authors had been trying to get working for several months. I found a few hundreds of "friend" declarations — any C++ programmer knows what it means: a few hundreds of global functions accessing private items, which really should not have been global. I doubted whether that software could still merit the title "object oriented," which might also explain the difficulties. Interestingly, those programmers did not seem to be innocent beginners; they were intelligent and they worked very hard.

The same problem relates to a specification. Many requirement details that become apparent during the coding phase flow directly into the code. In effect, at the end the code is the only "document" that contains the full description of the software. I was once engaged to write a software package that should have replaced the old one. Before I entered the company I was, of course, told that the software would be written according to well-defined requirements (the company external image must be intact). Actually, the code of the old software was the "requirement." It has been an interesting experience for me as:

I learned the difficulty of enciphering "what the code does" from the code — some passages were just impossible to understand, some were just dead code, probably doing nothing useful.

I learned how bad the code of old software, changed by innumerable people during several years, can be — a very interesting study of corrupted code that cannot be maintained anymore.

In Code We Trust

As a rule, maintenance costs increase with time. A decisive factor is the software quality in all its aspects: specification, design, and coding. Therefore, documentation plays a decisive role. It is obvious that the only document that corresponds fully to the running software is the source code. If the specification and design are in a good shape, they may contribute essentially to successful debugging or changes. Software developed without any (documented) method or concept reaches its limit earlier than software that can be also understood without studying the code. Unfortunately, that is seldom the case, which means that the basis of maintenance is as a rule the code. When programmers who wrote the code are no longer available, the maintenance costs become very high.

The other side of the maintenance problem is that we have to live with very old software, which is practically un-maintainable. At the end of the last century many experts expected on January 1, 2000 a collapse of software dependent on date. Those predictions created a demand for nearly forgotten languages like COBOL and enabled some already retired people to earn some money. Nobody knows whether the old software was really improved — probably in many cases no changes had been required but nobody knows the truth.

The Costs of Software Errors

It is difficult to make precise estimations of costs caused by software errors. In a NIST Planning Report 02-3 we have found the following data for 2002: the estimated costs of software errors equal $59.4 billions annually. The study says that although not all errors could be removed it is estimated that a third of them could be removed by improved testing infrastructure. The interesting but sad conclusion from reading that study is that the authors believe that only better testing can improve software quality. Obviously, they gave up the idea that better software development methods might produce better software.

The Programmers' World

People play a decisive role in software development. In spite of all development methods and tools, a team of programmers decides about the success or failure of the software projects. The dependency on human skill and motivation seems in the software industry to be more pronounced than in other industries.

The quality of any worker is determined by three factors: education, experience, and individual, personal ability. In contrast to other occupations, a programmer's individual abilities seem to play a more important role than education — at least it is a widespread opinion. Programming is a purely mental activity interrupted by coffee breaks, so it is difficult to compensate for missing abilities by extra devotion to routine tasks. A programmer cannot compensate for a bad concept by increasing the number of typed characters per second — the quality of code cannot be measured by the number of source code lines.

Sometimes we tend to believe that programming is a kind of artistic occupation. The problem with this view is that most artists have some sort of art education, generally working with their masters. Some people study nothing or anything and believe that it is a good basis to be an artist, or a programmer. Software development requires today so many people that for sure we do not have so many artists or specifically talented people who can develop good software without any formal education.

Therefore, more attention to useful education is a must if we want to improve software quality. Effectiveness of people differs but in software development the diversity exceeds any acceptable value — it is not uncommon to have programmers who are 20 or more times less productive than their colleagues. Considering such differences the question of "who is a programmer" seems to be legitimate.

The programmer's occupation is a difficult one. Eventually, they are responsible for the end product. Programming is a very stressful job, especially because software development comes at the end of a development chain. Only if the plant, machine, equipment, or devices are ready can software be installed and tested. Only when the software operates can the weak points of a technological (mechanical, electrical, chemical, etc.) design be seen. Unfortunately, programmers must very often first prove that the malfunctions of the controlled process result from the technological failures and are not software application errors. They are also forced to try to correct bad technological design decisions with software means, which is often impossible.

A paucity of good programmers and misunderstandings about programmers' qualifications are very important factors in the high price of software development, expensive maintenance, and poor software reliability.

A Programmer in a Project

For most programmers, the beginning of a new software project is a good and exciting time. The challenge is presented by the requirements, discussions about possible solutions, specification and design, the first lines of code. It seems that this time we get a chance to create truly good software. Unfortunately, in this beginning phase we lay the foundations for the problems we shall soon have — and sometimes for a disaster. We overlook too many things that seem to be negligible details, we do not care about proper documentation of the decisions, we leave too many questions without answers believing that they will be solved later in the implementation phase. We often start the implementation too early.

Once starting programming we are forced to mix solutions of coding details with application problems. While the programming continues, we are more and more often confronted with situations that were not foreseen, situations that require (impossible?) changes to software design, situations that force us to do dirty tricks that we "normally" would not do. Slowly, we get the feeling that we are losing control over the project. The worst thing is that we are as usual delayed although we were convinced in the beginning that "this time" our schedule is correct. In general, our mood is getting worse and worse, especially that the testing seems to be never ending and the pressure from managers and customers ever growing.

Discovered and removed errors produce nearly the same number of new problems. At the end of the project we are totally frustrated and, independently of the result, we want only one thing — to get away. We have asked ourselves for the reasons and do not know the answer.

That scenario is valid even for software projects that are relatively well planned with reasonable schedules and financing. Badly planned software projects and especially their maintenance are just nightmares that require programmers with exceptionally good mental strength.

The Software Project Leader

At the end we would like to comment on the role of managers in software development, but we limit our remarks to the lowest level: software project leaders, the upper management levels being outside the scope of our considerations. Because of software specifics it seems that those leaders must have a strong programmer's background. Good programmers do not like to be managers but some programmers must do the job. Leading a software project cannot be a purely administrative job. Therefore, even people with a good organizational talent will fail because of the ignorance barrier between them and their team. Unfortunately, there are too many innocent, ignorant project leaders who make their contribution to software disasters. They do it by accepting unrealistic wishes on the one side and not understanding the true problems of software development on the other side.

Examples of Disasters

The history of software provides enough examples of projects that were interrupted because there was no chance to reach the goals — in the software case the entire investment gets lost, there is usually no way to reuse the partly done code in other projects. We do not directly discuss here examples of the better-known software catastrophes but for the reader's convenience we provide a few comments about unfortunate software. A search for "software disasters" on the Web results in thousands of publications that disclose details of software projects that failed and led to financial losses or even killed people.

The consequences of software failure are especially "unpleasant" in space flights because they result in a damage to the rather expensive shuttle and instruments and all that under the critical public interest. Ariane 5 explosion, Mars Climate Orbiter loss, Voyage II, Marine I, Phobos 1 and 2 are better known examples of catastrophes in space due to software errors. Nobody knows precisely how many research projects in space have been canceled or not done as planned due to software errors.

The automated baggage system for the international airport in Denver could not be realized on time on account of software design difficulties. Because of the huge financial losses it was a favorite topic in newspapers and on TV.

There is probably no domain of human life that has not suffered from software failures. Telecommunication (service fails for several hours), airplane and railway disasters, ticket reservation software, banking software, nuclear reactors, election machines, ATM machines, power supply (blackout), construction disasters (bridges, buildings), all have experienced software that does not work properly.

Military software is probably a very special domain where many people have lost their lives (friendly fire as an example) due to software errors — secrecy of military issues covers many such disasters with silence.

As we have learned not to trust software solutions, we do not transfer to it vital safety supervision which may cause injuries. This explains why software failures are rarely responsible for killing people. A better-known example can be found in Baber's paper,[2] which describes a case where patients were killed by a cancer treatment machine, the Therac-25, whose faulty software on rare occasions delivered massive overdoses of radiation. That paper contains a good analysis and critical arguments about software engineering. It also describes several other cases of software disasters.

A number of software projects are eventually successful, leading to working products, but financially a disaster (e.g., the truck toll collection project in Germany described by Borchers[3] but a government can pour any amount of money into the sand).

Recently, I bought a DVD recorder, which is an interesting and sad example of the "disasters" in the Hi-Fi industry. That industry has been resistant for a long time to software solutions, but eventually it has been forced to digitize all its equipment, effectively doing most tasks, especially relating to the user interface, by software. The device I have bought is incredibly bad: the user interface is terrible and I can crash it at any time. Are we so used to bad software that we should accept it even in devices that are intended for people who are not software freaks and are not really interested in investigating which key sequences will lead to success? The situation will not change soon, as the Hi-Fi industry does not have any software culture and it will take some time before the industry realizes that it is possible to write better and more reliable user interfaces.

There are some little-known software projects that are done according to a timetable and that do not exceed the planned expenses. Many companies start software projects assuming that the plans are unrealistic but the reality very often exceeds the managers' worst nightmares. Statistics show well the extent of the problem. For instance, in a Paulson[4] publication,* we found the following numbers:

> The report stated that there were 78,000 software development projects judged as having been successful during the year compared to 137,000 that were late and/or that exceeded their budget. Another 65,000 projects failed.... Cost overruns they incurred were at 45 percent in 2000; time overruns at 63 percent.

Recommended Reading

1. Brooks, F. P., *The Mythical Man Month: Essays on Software Engineering.* Reading, MA: Addison-Wesley, 1995.
2. Baber, R. L., "Comparison of electrical 'engineering' of Heaviside's times and software 'engineering' of our times," *IEEE Annals of the History of Computing* 19, no. 4 (1997): 5–17.
3. Borchers, D., "Verursacherbedingt verspätet," *c't*, no. 22 (2003): 92–94.
4. Paulson, L. D., "Adapting methodologies for doing software right," *IT Pro* (July/August 2001): 13–15.

---

* By the way, that paper contains a very convincing plea for more methodology in software development.

Chapter 3

Software as Engineering?

Methods

There are some difficulties in discussing the term "software engineering."
If we understand under that term the body of knowledge about creating
programs, software engineering should cover:

- Specification methods
- Designing methods
- Programming techniques

To round it up, software engineering should also contain:

- Organizational topics

The reality differs from those expectations. As observed in several publications, software engineering does not contain knowledge about *engineering software, understood as knowledge about how to write a good code.*

Actually, Agile methods would have completely lost their justification
if coding had not contained a design element. Therefore, it is possible to
write code without a formal design. Whether it is a practice to be
recommended is another issue. Whether it is possible to code without
any design work (i.e., to serve as a pure translator converting a design
into a code) is also doubtful — if it were possible, coding could be
automated and the programmer would be replaced by translation programs. Interestingly, the concentration on management and organizational

topics while neglecting technical aspects has been criticized for several years without visible reaction. For example, we find a very constructive criticism of software engineering done by Baber,[1] and similar arguments can be found a few years later in Whittaker et al.[2] The common (mis)understanding of software engineering is at best formulated in the abstract of the Whittaker et al. paper[2] mentioned above*:

> Books on the subject favor the "light" side of the discipline: project management, software process improvement, schedule and cost estimation and so forth. The real technology necessary to actually build software is often described abstractly, given as obvious or ignored altogether.

The reason that books on software engineering bypass the technical problems of programming is that the problems are very difficult. For instance, where can a programmer find indications about definition of base classes or in general data structures applicable to a given application? There are many questions: How to get a proper balance in validation of variables? How to solve the client–server relationship or more specifically how to choose between inheritance and composition? How to handle errors? When and how to use exception handling? How to use effectively runtime libraries? How to find a proper way in the jungle of system calls? How to program parallel processes? Those kinds of questions are not about the syntax of the programming language: the successful use of any programming language requires a specific software engineering approach. To answer them we also have to be good programmers. Maybe good programmers do not have time or cannot write books. The programming world knows two answers for missing software engineering topics: very good programmers and trial-and-error methods.

Very good programmers are very rare and some of them believe that they do not need software engineering. Therefore, if a very good programmer leaves a company the remaining persons may discover a desert in place of the wonderful software and they have to start from scratch. Most programmers are not brilliant — they are just normal human beings who have to compensate for missing knowledge by just hard work: they try, they "search the Internet for tips," they try again, they discover the wheel anew, and after a while they find a solution for a tiny piece of their problems. The solution might not be the best one but they are happy that it works — the outside world cannot in any way judge what is going on in their code.

---

* In general, we can recommend the paper to anybody as it contains a very good characterization of software practice and actual software problems.

Agile methods are just a set of rules on how to organize a software project. We understand that it is a problem that should be properly solved, but we know also that programmers' skill plays a decisive role in any software project. A few excellent programmers will beat any well organized but essentially larger team containing average or weak programmers. Organization can compensate only partly for missing skill in true software engineering.

As the essence of the topic "how to write a good program" is so difficult, the experts bypass the difficulty in their books, producing pseudo-scientific methods or concepts that do not contribute to a better software world.

Fascination with Graphics

The favorite method of dazzling innocent observers is to define some fine-looking graphical symbols, add fine sounding names, and declare it a method. In this way, standards obviously produced by some administrative craftsmen fake a quasi-scientific approach.

What is actually a method? We would expect that it is a technique based on possibly algorithmic steps that can be proved or at least verified, and preferably based on some mathematical principles. A method is defined by a set of rules.

PLC programming recently obtained a standard (IEC 61131-3), which is a good thing in principle as it forces manufacturers to leave their proprietary niches and open their products. On the other hand the components of that standard demonstrate how "methods" or "concepts" are understood in parts of the modern world. We use symbols, e.g., rectangles, to explain something. For example, we draw two rectangles, link them with a line with an arrow and say that the output of the first rectangle is passed to the input of the second rectangle. It is a common way to explain something to another person, as it is much more comprehensible than a purely verbal description. But we would not then claim that we defined a method or concept. That is exactly what happens in several standards. The same rectangle named as a Functional Block used on a drawing named Functional Block Diagram defines suddenly "a powerful concept which encourages the development of well structured software, ... it can be reused ... and it has many properties found in object-oriented design." It is just incredible, reading that we could really have doubts whether we understand the concept of the object-oriented approach. We are used to such flowery language in marketing prospects but if it is seen in (pseudo-) scientific papers and even in IEC Specifications it is not amusing; it is frightening. (See Appendix E *Going beyond the limitations of IEC 61131-3* for further remarks on this topic.)

Visual Basic Chaos

Speaking today about BASIC we mean in most cases Visual
Basic. In Visual Basic it is very easy to create a graphical
interface using the typical GUI symbols such as buttons, text
boxes, check boxes, combo boxes, tree views, list views, etc.
This makes Visual Basic attractive especially for a nonprogram-
mer who wants for some reason to program. Usually, a set of
graphic symbols does not make a program. Therefore, each
graphic symbol possesses a number of procedures linked to
events generated by the symbols, such as click (button), change
(text), select (list), etc. All this is well organized, clicking on a
symbol opened a source code editor with predefined procedure
name, parameters — all very nice and convenient. What is truly
missing is a method of how to write the code or, better, how
to design or to at least organize the software.

We know of course what to do when we click on a button or select
an element in a list. But in any nontrivial software those activities are not
totally separated and time independent: they are coupled together and
activities triggered by events are dependent on "states" of other graphical
symbols, in general on the present situation. As people are intelligent,
they find a solution, but the results are terrible: all based on global variables
manipulated by any procedure in a way that nobody can understand,
after a while even the author of the code.

Those ways of creating software can be found in several software
packages, which are proudly described as "tools which allow building
complex control systems" by creating a diagram of the application using
predefined graphical symbols. Those descriptions imply that the true
problem — the control — is just a secondary problem: it will be arranged
in some way behind the nice graphical façade. That assumption is just
nonsense: behind the graphics we find an uncontrolled chaos created in
hours and hours of laborious work where the main activity is to find out
why the flags or markers, whatever their name, are behaving in such a
way and not as we expected. The reason is obvious: the software has no
design; the required behavior of the application has been transformed
from a verbal description into a set of intuitively linked procedures. We
have to accept that common sense has limits, beyond which it produces
only chaos. The argument that it often works is not convincing: working
chaos is still chaos.

Object-Oriented Design Illusion

A missing design concept or at least a lack of coherent structure of software also can be seen in other approaches even if they take proud names of, e.g., object-oriented design. Object-oriented design is a good method but it has its limits — it is not a solution to all software problems. Encapsulation of data and their affiliated methods is a truly powerful way of creating software. But object-oriented design does not provide the answer to the question: how to organize a set of objects. Each object is able to perform a well-defined function on stored data and any other object can claim that functionality. But we do not know even how to use inheritance and association effectively as the basic "links" among objects. Advice like: "after many years of experience we know that we should not overdo the use of inheritance" is good but still it is hardly a method. Defining classes graphically does not change the situation either: it is only a different way of definition.

UML Illusion

UML is another example of a notation that is hugely overestimated. If we define a class in code we consider it a declaration of a software type. The same class defined in UML and appearing as a graphical symbol on some diagram is also just a definition of a type. Or is it more? To convince us that we have to do the design in UML (although we know that we will need to repeat it anyway in the code) we must have convincing arguments that we can do more on the UML diagram than in the code. Unless we have the arguments we may express at least critical doubts about usefulness of such notations. So, several authors express their doubts about the alleged advantages of using UML. A fairly representative picture emerges, putting together, e.g., the remarks summarized below.

Glass expressed his doubts[3] about "theoreticians' view of software modeling" citing among others that such a view:

> ... is clear that generating code from the evolving problem model is unlikely to work.

Mellor, a proponent of UML, finds only the following answer to that reproach, an answer that states that UML is just a text that a programmer translates into a code:

> How comfortable are you with the idea that you could build a textual representation of a state machine and use it to choose which function to execute next in response to some signal?

In that context, in another paper, Douglas[4] from iLogic (a company that sells Statecharts) admits:

> You cannot translate Statecharts directly into a design-object model.

In spite of those comments and statements, various UML marketing specialists "sell" it as a method for creating executable specifications that are nearly automatically translated into code.

We do not want to condemn UML; as a common modeling language it is an interesting idea, but UML contains so many "nice" points that are representative of pseudo-concepts and methods that we take another example: the belief in the power of words that are becoming sacred cows as if they defined something excitingly new that revolutionizes the software world. Let's take the term "Use Cases," which is another name for "sunny-day-scenarios" (sunny-day-scenarios describe expected situations and therefore is a clear term). Of course, Use Cases are a skeleton of any application and in some applications (business software) they may be the dominating situations. But even in a case of selling or buying something, which are the favorite examples in teaching UML, we should have a chance to express situations when things go wrong. Sunny-day-scenarios normally do not present any challenge, the true complexity of any software results from the erroneous, unexpected situations. Taking a non-business example — a motor control, which should be switched on and off via a motor controller. The expected actions (Use Cases): switch motor on and switch motor off are just peanuts; to express them we do not need any methods or design. The true difficulties are represented by some easily imagined problems:

- The movement command parameters are not accepted by the motor controller.
- After sending the start command the motor does not start. Possible reasons: perhaps the device is already in the required position, the interface does not work, the controller is down, the power supply is down, the cable is broken, etc.
- The motor stops before reaching the required position; perhaps the motor is overloaded, the motor is too hot, a limit switch has been encountered, the break switch has been pressed (which bypasses software), etc.
- The motor stops and signals erroneously "movement done" although it stays in an incorrect position.
- The motor controller never signals completion of the movement.

- The supervisory software demands to stop the movement (the user pressed a stop button, a cooperating device requires to break the movement, etc.).
- The motor does not react to the Stop command.

Obviously, a UML designer must, from the beginning, assume that the true problems will be solved in the code and his specification handles only a "higher level" of behavior. He uses UML as a way of formalizing the requirements document for use in the programming phase, but the usefulness of specifications that ignore the genuine application problems is doubtful.

Formal Methods

In parallel to pseudo-methods there are serious approaches based on sound theoretical grounds. Unfortunately, they are done only at universities and totally ignored by industry. Reading scientific journals we get the impression that nearly all software topics are discussed and solutions are proposed for several problems. But there is no link between universities with their scientific papers and industry. Universities and industry are two separate worlds trying to solve the same problems but in their essence following different goals.

As state machines are the essential topic of the book, it is interesting to compare their use and presentation in the two worlds. In the scientific world the parser definition of the state machine dominates in software application. Discussing hardware design, scientific papers concentrate on model definitions, optimization of state number, verification methods. All those theoretical topics are of little practical usefulness and do not make too much sense in the design of an industrial control system. Hence, the knowledge and the use of state machines in industry are half-hearted, and accompanied by several misunderstandings due to lack of a sound theoretical basis.

CASE Tools — Value for Money?

> The central problem for any true progress in software development is the replacement of programming as understood today: translating complex requirements using the relatively primitive means offered by even the best programming language.

Replacement does not mean elimination. Our present understanding of software leaves large fields that still require the old, proved, and successful methods independently of our feeling of discomfort with them. By replacement we mean that the dominance of the established method must be weakened. The way to go is to strengthen specification not as an intermediate development step but by conceiving of a specification that is executable; this will eliminate coding in wide areas of software development.

The problems encountered in introduction of specification methods and tools in programmers' real life have been discussed in the previous chapters. In fact, the case is rather simple:

> Specification tools will start to play a role in software practice if the benefits exceed the time and money invested.

The investment is greater than the pure costs of the CASE software. The true expenses result from training and required adjustments in software projects. Training cannot be neglected; especially in case of specification methods, the topics must cover not only the use of a development tool but also the teaching of methods. For instance, persons without any knowledge about automata theory or control engineering will have basic problems in understanding finite state machines or Petri nets if those concepts form the basis of a specification method.

Programming or Specification Languages?

Controversy between the followers of specification and the supporters of coding has some more philosophical aspects. Usually, a programming language is well defined as a set of instructions for a computer, which if properly arranged forces the computer to do something useful (or at least does not crash it). A definition of a specification (as a language) is hazier. A specification is more than a verbal description; it is a formal description. "Formal" suggests here that the specification may be verified and clearly transformed into some other presentation (code). If the specification is executable — does the specification "language" become a programming language? If the answer is positive — any programming language can be considered a specification language. In fact, this is the true background of the Agile approach: apostles of that claim that they "specify" the requirements directly in the programming language. Thus, even some fundamental issues may be questionable and depend on just definition. Jokes aside, software engineering technology is really difficult, which explains why there is so little done in the principal topics: how to write good code, which might be translated into how to make an executable specification.

Development Cycle

Actually, such topics as project management, software planning, schedules, and costs estimation are quite well investigated and presented in publications. Those topics are outside our allotted scope but we would like to look briefly at one specific issue: the software development life cycle. That topic covers the understanding and definition of software project phases and their influence on software reliability and costs.

Software "life" is limited: software is developed, used, and after some time "dies" as no longer usable. There are several reasons for software death: it is old-fashioned, demand disappears, it is impossible to adjust to new requirements. The software life-cycle consists of several phases. In the course of time several models of the software development life-cycle have been studied: waterfall, spiral, code and fix, rapid prototyping, etc.

The software development life-cycle models can be quite complex. The reasons are obvious if we look at the factors that could be taken into consideration. From the pure software development point of view there are such topics as Proposal, Feasibility study, Requirements, Specification, Design, Coding. Testing cannot be forgotten — here we have Modules testing, Testing while programming, System test, Acceptance test. The result of the software development suggests topics like Prototype, Pilot system, Final system. The software quality could be represented as Code quality, Software security, Software reliability, Documentation. Considering the user, we should take into account: Deployment, Installation, Manuals, Training, Maintenance. The list is certainly not exhaustive, but at least those topics play some role in discussing a model of software development life-cycle.

There are several models of software development life-cycle analyzing or defining each step in a software project. There are no universal models. Further, not all software projects require all imaginable steps. For instance, let us discuss two activities that are contradictory: Prototyping and Specification.

Prototyping

Prototyping means to build relatively early in the software development the first version of the application. Prototyping assumes that the application cannot be well specified. Hence, the proper way is to develop something quickly and start testing, including in that process as many participants as possible. In addition to the programmer, anybody involved in some way in the project can and should be present to assess and to influence the software: colleagues, managers, marketing people, customers. Prototyping is an iterative process where the application emerges as a result of several trials. That idea (close to Agile methods) makes sense in certain applications where experience shows that it is impossible for the end user to specify the software. Classical examples of that software are applications where a GUI is just the application or plays a dominant role. For instance, for a web design it is impossible to specify a good layout without seeing and trying it. Only the cooperation of several persons who criticize the GUI can produce a good application, which is not only nice looking but also ergonomic.

A good prototyping process assumes that this is not the final software. Unfortunately, that is only the theory. In practice, it is very difficult to consider a prototype an exercise that is then a basis for the true software development. Therefore, applications that start with a prototype very probably end with terrible code. Managers, especially marketing people, like prototyping because the first results are quickly seen and they hope to sell the prototype as soon as possible.

Specification

There are applications that must be specified — they do not need a prototype in the sense of the word as discussed above. We recall that a prototype is needed if there is no common agreement about how an application should look or behave and that agreement cannot be achieved by some theoretical or artistic considerations. Applications in telecommunications are precisely defined by standards. Similarly, applications in industrial control are completely defined by various control and safety requirements. Even if we suspect that it is impossible to make a perfect specification, we specify the software as well as we can, based on available documents, standards, requirements. Those are examples of applications that are well defined and there is no need to "discover" or define their required behavior by tests and trials. The specification errors resulting from overlooking or misinterpreting documents will be corrected by test of the created software. In such a case we start detailed design and coding as late as possible when we are convinced that the specification is ready and we can develop the final software.

Software Development Steps

Figure 3.1 shows the most important steps in software development. It is reminiscent of the waterfall model — but only the changes arising during Coding and later by Testing and by Maintenance are shown. The theoretical model in Figure 3.1a assumes that those changes influence Requirements, Specification, and Design and that the corresponding documents will be updated. Usual practice (Figure 3.1b) differs from the theory — as a rule

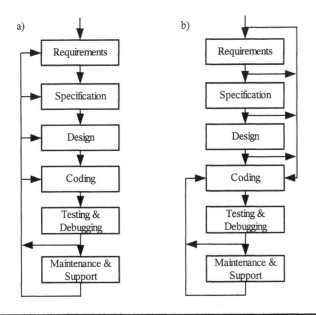

Figure 3.1   Software development steps: (a) in theory; (b) in practice.

the changes flow directly into the code, which is the center of all activities once programming starts.

Software Documentation

In each development step some documents are created — documents that are the entry point of the next step:

- Requirements: a verbal description containing at least a summary of required features should be prepared by a customer and agreed by the software company.
- Specification: a formal description of the application; as a rule prepared by a software team or specialized "specification" group.
- Design: a document describing the software project (the basic "layout" of the program and development tools used), sometimes also design reviews; prepared by the software team.
- Coding: source code and protocols of code reviews; done by the software team.
- Testing and Debugging: protocols of found and corrected errors; done by the software team or specialized testers.
- Maintenance and Support: protocols of changes, found and corrected errors; done by the software team.

Sometime during the development process the software documentation is created. We do not provide on the diagram a special step for documentation as it is created all the time. In fact, the results of all steps in the life-cycle could be the documentation. Unfortunately, reality is as usual cruel, and it is very difficult if not impossible to create a development environment in which one document passes smoothly to another one. It is less a problem of formal layout of the document. The core of the problem is the inconsistency of documents resulting from a sequence of development steps as shown in Figure 3.1b, and explains why a part of the software functionality is not present in any formal document and stays buried somewhere in the code. In my programmer's career I saw plenty of (pseudo-) documents that were supposed to be software documentation but in reality only the layout and title were appropriate. There is little motivation to update documents that were written earlier in the software development life-cycle; actually, the only "document" that corresponds exactly to the working software and is as such always up-to-date is the source code. Even this statement must be treated with some reservation: the code is of course up-to-date but some comments may be wrong as they are ignored by the compiler.

Testing and Debugging

Testing and debugging are similar but actually they should be done by different persons. Programmers who have done the programming work, who have been also involved in the design and maybe in the specification have difficulty leaving the "known" paths. Programmers test software during coding — it is an inherent part of coding and they develop habits to always repeat the same tests, which means that only a certain part of software is quite well tested. Only an outsider may find errors that are outside that programmer's routine. Definition of test scenarios is dependent on testers — there are some persons who have a "good hand" to crash any software. In spite of all efforts some bugs stay hidden because it is impossible to test all paths through the software.

Results of tests are diagnosed errors. To diagnose the causes and to remove them is again a job for programmers who develop the software. Hence, testing and debugging require a good cooperation between two different groups of persons.

Diagnosing errors is in many cases a very difficult job. There are several tools that are supposed to help programmers in testing. Especially, IDEs offer here very powerful means. In spite of all tools, testing is done very often on a very basic, primitive level. A search for "printf" statements, including those commented out, in a program may deliver an interesting result, giving insight into a programmer's mentality or dispelling illusions of powerful development tools.

Maintenance and Support

The last phase in software development lifetime may last a very long time. We gave it a double name to underline two different activities during that period: Maintenance means removing code errors complained of by customers, while Support covers software functionality changes demanded by customers. As software is going to be changed during that period that possibility should be taken into consideration during requirements, specification, design, and coding phases. A lack of attention to software maintainability leads to excessive software maintenance costs.

It is rather difficult to define any specific technical rules for software design that should make software maintainable. Software maintainability is assured by its clear structure, complete documentation, and well-defined automated testing procedures. Those factors are especially important in situations where the maintenance will be done by persons who have not done the development. The probability of that increases with time.

Human Factors

People cannot be neglected either. I had an interesting experience in that field. Several years ago as a consultant I introduced the Vfsm idea into AT&T. The full implementation of the idea took several years and Flora-Holmquist et al., who had participated, wrote a paper[5] that exposes the chilly reception of the potential users in that company. They defined five schools of philosophical thinking for maintenance of the status quo:

- But our [feature/application/process/state machine/code ...] is different.
- Is it just the paradigm of the month?
- Everybody already knows C, so why change?
- Show me the data.
- An FSM modeling environment; a novel idea ?... NOT novel at all.

We do not want to go into the details of those pseudo-arguments — they can be found in the cited publication. Several variants of that kind of resistance can be found in any environment — it is just life.

Summary

In this chapter we have touched on a few topics in the software development life-cycle. Although those topics are rather loosely coupled with the mainstream of this book, they deliver additional arguments for the complexity of software development. The complexity is not only of technical nature, but it refers to organization of software projects, required development steps, and indicates the role of all persons involved in the development process: managers, people who specify the requirements, designers, programmers, testers, and last but not least the customers. The role played by people in software development, especially their contradictory characters represented in extreme cases by "free artists" on the one side and pedantic bureaucrats on the other side, reflects the important role played by nontechnical topics in software engineering. But this does not justify the relatively modest representation of "programming" matters in software engineering.

Education Requirements

Although the non-programmers' topics are relatively well represented in publications and education of software engineering they obviously have little impact on software development. We need a more technically oriented software engineering curriculum. In addition to very useful disciplines like: project management, schedule, cost estimations, and the like, we need topics that discuss how to construct (design and code) a good program. The entire body of software engineering cannot be a purely theoretical matter: creating a program is a complex task, which cannot be solved by theoretical considerations only.

Who Is a Programmer?

Software development requires several types of skill. It begins with the capability of coding through basics of informatics and ends with specification methods. It is true that the one thing that a programmer must definitely know is the syntax and semantics of the programming language used. But it is too primitive to think that such knowledge of "how to code" makes a programmer. The more advanced topics are as important as the elementary programming language information. Therefore, we formulate a thesis that a programmer is a person who acquired the educational chain:

syntax — basic of informatics — specification

that is, the programmer is a person who is able to specify the software, design it, and code according to the present state of the art in information technology.

Education as the Basis of Skill

The rapid changes in software and the expansion of the software industry created a huge demand for programmers. Therefore, this occupation is practiced by any willing persons who can experience a "joy of coding" feeling. In the early years of software development their quality did not play a very important role — the desire to contribute compensated for missing skills. As the software industry matures and its products become more and more complex, the education of programmers is becoming an essential factor that will decide the success or failure of a software project. Examples of individuals who are remarkably good programmers, with or without any formal education, are misleading. They must be complemented by millions of average programmers whose best weapon is a good education covering specification and design methods combined with effective coding.

Missing Skill — Examples

Speaking about methods, we have already hinted at the broad bend the authors of software engineering books do around technical topics. Their favorite subjects are just detailed discussions of language syntax, in other words explanations of the heavy language of standards using more informal concepts supported by examples. This is necessary and useful, but it is far from adequate. Especially for those who are not completely greenhorns most books are useless because they are looking for answers to more advanced questions. Those advanced questions are just mentioned or compressed in a short chapter at the end of the book or lectures as if they were to be learned by doing or were obvious extensions to the basic knowledge. Unfortunately, it is not the case — advanced topics are the essence of software design and determine the success of a software project.

To the general topic: as a design of a program is very difficult we limit our comments to a few examples of specific topics, which can be seen in most software projects, to demonstrating the missing understanding of some basic software concepts. This applies to all concepts that are not as a rule part of the programming language, e.g., exceptions, events, threads.

Events and threads are among the most misunderstood and overused concepts in software. The reason is that programmers are not taught the sense of using them. Concepts that are essential for some software such as operating systems or input/output drivers may be totally useless for application software. We have made in this book some spiteful remarks about PLC but they also have a very important feature, which explains their long and successful history — their guaranteed response time due to the polling, which is the basic mode of PLC operation (to be fair we should also mention that there are customers who justify the use of PLC by the guaranteed response time, although they do not need it). So, when speaking about threads and events we have to know what their disadvantages in comparison with polling are, where we need threads and where they are just wrong. A programmer must also understand what it means to have hundreds of threads waiting for events and sending messages — such a system can function only when very carefully designed.

Exceptions are a controversial topic. The original concept was to catch software errors, i.e., errors resulting from failures made while programming. Opponents argue that we should explicitly catch those kinds of errors in code. We do not want to try to solve the dispute, pointing out only to situations like inputs from outside that may cause problems due to their random character and are not classical programming errors (it is difficult to foresee all imaginable behavior of the outside world). But we encounter software that uses exceptions to solve all application errors and we would classify it as a total overuse of that concept. In other words in such a software only the sunny-day-scenarios are programmed. Any applications errors are left to be handled by exception (if the temperature is too high the exception is thrown). It seems that authors of such software are missing some basic understanding of software engineering (or that we have a wrong definition of it).

We should also mention the overuse of pointers practiced by some programmers (a specific problem of C programming). Perhaps the reason is just that pointers are well presented in the basic programmers' education, and therefore for some programmers they must be used in any situation (e.g., as function parameters). It has not only consequences for heap and stack but is a general problem of programming culture.

With those few examples, we wanted once more to support our disappointment about the overweight of management topics in comparison with technical ones. A superior project management will never compensate for unskilled programmers but the contrary statement is true: that bad management will not prevent good programmers from writing a reasonably good program.

Programming skill understood as joggling with basic structures of the programming language is not enough. Programmers require a better education, which goes far beyond that and covers advanced topics that allow them to design a program with care, responsibility, and deep understanding of the power offered by programming languages, libraries, and other development means.

Conclusions

Here we summarize Part I, presenting first different aspects of the discussed software problems. At the beginning of this part we made clear that we present here our views and not the results of some formal study. The view is based on our experience and knowledge collected during several years work in the software industry. We are not alone in our assessment of software development methods. We have seen hundreds of publications about software development and read many of them. We have made a few references to them but they do not represent any carefully investigated and representative choice of publications in the software domain — they are just examples of publications to which we have access and which we like. Those examples prove that we are not alone or exotic in our opinions about the poor state of the software industry, as characterized by increasing requirements that are not matched by true progress in development methods.

If we understand under the name *software engineering* all topics related to "how to create a program," we diagnose that software engineering is in an unbalanced state: the difficult technical questions are not answered. We know some quite good solutions for management of software projects but we miss convincing methods for designing and writing good code.

We round out our consideration with a suggestion of a new way to go. But before we formulate our proposal we summarize first the reasons for software problems. Let us begin with some trivial truths, which probably should be the entry point of any discussion about software

- Programs are complex.
- Code is impenetrable, especially when written by somebody else.
- Unless we are producing something that can be executed and/or analyzed — we are *not* doing anything useful.*

Those contradictory statements show the entire extent of the problem. Having those things in mind we understand why the controversy between the extreme positions in software development methods: *specify first* or *start with coding* (see the section *Specifying or Not?* in Chapter 1) is very difficult to resolve.

We have criticized some trends in software development that are pseudo-activities, expressing helplessness rather than a true ability to offer solutions. There are many pseudo-solutions that may have some value in certain circumstances but in general provide little help. To that category belongs not only the overemphasized role of UML but also several "common wisdom" based opinions as for instance:

---

* We have adapted that sentence from a reader's letter (J. Chludzinski in a discussion initiated in Embedded.com by a paper Coders vs. Programmers by J. Ganssle). The letter relates to a specific topic but we believe that it is at the core of software problems.

- A belief that human intelligence can compensate for a missing skill.
- Fascination with programming languages, which leads to overproduction of similar languages, effectively deepening the difficulty, at least from the educational point of view.
- Fascination with graphics and considering any graphical representation a method.
- Creating reliability by testing.

Those reflections suggest that there is no clear answer for software dilemmas, in other words:

> There are no general methods or solutions for all software.

We are convinced that the software industry must leave the established ways determined by marketing and try another way. That other way is the topic of the following parts of the book. That way assumes that software can be based on a ready-made execution environment, which is written once and then used to build many applications. The adaptation to a given application is achieved by establishing a specification of the application's behavior and its configuration, with the results of that specification carried out by the standard execution environment. This approach means a greatly reduced coding effort, replaced by specification. That approach works only if the specification methods and development tools allow a totally complete specification (to the last details) to be generated. That approach reduces the dependencies on program complexity. That approach operates with a specification that is executable and that is not in an intermediate form but is the final application implementation.

Recommended Reading

1. Baber, R.L., "Comparison of electrical 'engineering' of Heaviside's times and software 'engineering' of our times," *IEEE Annals of the History of Computing* 19, no. 4 (1997): 5–17.
2. Whittaker, J.A., Atkin, S., "Software engineering is not enough," *IEEE Software* (July/August 2002).
3. Glass, R.L., "On modeling and discomfort," *IEEE Software* (March/April 2004): 102–104.
4. Douglas, B.P., "UML for executable specification," *EDN* (August 2001).
5. Flora-Holmquist, A., Staskauskas, M., "Moving formal methods into practice: The VFSM experience," *Proc. of FSMP'96, 1st Workshop on Formal Methods in Software Practice* (1996).

Part II

FINITE STATE
MACHINES

Chapter 4

Introduction, Definitions, and Notation

Finite State Machine

Consider a system as shown in Figure 4.1 where a Control System controls an Application.

The Control System receives a number of stimuli (Inputs) from the Application and produces actions (Outputs) to affect the application. The control system realizes the control using simple logical conditions of the form:

```
If (Input Conditions) Then Outputs
```

The input conditions are logical expressions formulated according to Boolean algebra. Some examples of Input conditions:

```
valve_open
```

```
valve_open OR timer_expired
```

```
valve_closed AND temperature_ok
```

where OR and AND are logical operators. All arguments in these expressions are Boolean values (false, true). In the approach presented in the book we will expand the range of values in input conditions beyond the Boolean values and therefore the NOT operator will not be used.

This simple model is sufficient for rather trivial systems. Any more realistic applications require a much more sophisticated control model. One of the most powerful models is a **finite state machine (fsm)**, which is used to describe behavior: what to do in all imaginable situations.

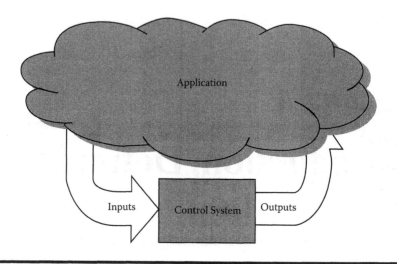

Figure 4.1   A control system and the application.

The finite state machine introduces a concept of a state as information about its past history. All states represent all possible situations in which the state machine may ever be. Hence, it contains a kind of memory: how the state machine can have reached the present situation. As the application runs the state changes from time to time, and outputs may depend on the current state as well as on the inputs.

As the number of distinguishable situations for a given state machine is finite, the number of states is finite. This explains the word "finite" in the name. Because we discuss only "finite" state machines in this book, we abbreviate it to "**state machine.**" The term "finite automaton" is also widely used as a name for this concept. The fsm concept is defined and discussed in innumerable books and papers. We cite only two good publications in the references: the Gomez paper[1] and the information presented on the Web.[2]

A control system determines its outputs depending on its inputs. If the present input values are sufficient to determine the outputs, the system is a **combinational system**, and does not need the concept of state. If the control system needs some additional information about the sequence of input changes to determine the outputs, the system is a **sequential system**. The logical part of the system responsible for the system behavior is called a state machine. Sometimes combinational systems are treated separately, and sometimes they are considered a kind of degenerated state machine, with only one possible state, to which we could give the name "*Always.*" To keep things simple, we call any logic that determines system behavior a state machine. According to this definition, a state machine can be represented by the diagram in Figure 4.2. The history of input

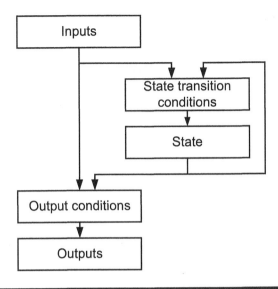

Figure 4.2 The state machine definition.

changes required for clear determination of the state machine behavior is stored in an internal variable **State**. Both the State transition conditions and the Output conditions are functions of Inputs and a State.

A control system realized as a single state machine is of limited usefulness. To stay comprehensible a state machine cannot be too complex — the numbers of inputs, outputs, and states have limits. These limits are not fixed constants but result from common sense and depend on the application, the designer, and other factors. For instance, from the documentation point of view a state machine transition diagram should fit on a single A4 or A3 (or corresponding U.S. sizes: letter or tabloid) sheet of paper. Therefore, serious applications can be realized only by using several state machines, which should be organized in a system according to some predefined rules. Otherwise, one could end up with a state transition diagram covering entire walls. This is a vitally important point, which is further discussed in Chapter 9 *Systems of state machines* and elsewhere.

State Machine Models and Presentations

Transition Matrix

A state machine may have different forms that define its ability to describe behavior.

The simplest state machine model is a Parser. The function of the Parser is to reflect input changes. The Parser does not produce any specific Outputs — its state is the output. A well-known example of a Parser is a state machine that recognizes certain strings. The functioning of a Parser may be shown using a transition matrix or a state transition diagram.

Example

A control system has to count the amount of money dropped into a vending machine. To keep the example simple let's restrict the inputs: only 5 and 10 cent coins are accepted. The correct, recognized sum is 25 cents.

A **transition matrix** is a table where rows (*From*) represent present states and columns (*To*) — next states. The content of the table are conditions that must be fulfilled to make the transition from a present state to the next state true. The transition matrix for the example is shown in Table 4.1.

The input conditions are numbers 5 and 10, which correspond to the recognizable coins. A missing condition (-) means an irrelevant or inapplicable situation (cannot happen). The state *Start* is the initial start: there are no coins. The state *Stop* means: there are 25 cents dropped. In all states but *Twenty* both coins are accepted. In the state *Twenty*, the 10 cent coin is ignored and rejected, although different strategies could be considered in practice.

A Parser is a state machine that has a beginning state *Start* and a last state *Stop*. When a Parser reaches the state *Stop* its task is successfully terminated and it must be reset to carry out the next recognition process. This sort of machine produces a definite result by means of a single run through a procedure.

The alternative presentation of the transition matrix is shown in Table 4.2. In this table the columns represent inputs and the rows — states or vice versa. The content of the table are next states. We use the form that is more convenient for a given application.

Table 4.1   Transition Matrix for the Vending Machine Counter

From \ To	Start	Five	Ten	Fifteen	Twenty	Stop
Start	—	5	10	—	—	—
Five	—	—	5	10	—	—
Ten	—	—	—	5	10	—
Fifteen	—	—	—	—	5	10
Twenty	—	—	—	—	—	5
Stop	—	—	—	—	—	—

Table 4.2   Alternative Form of the Transition Matrix

Input State	5	10
Start	Five	Ten
Five	Ten	Fifteen
Ten	Fifteen	Twenty
Fifteen	Twenty	Stop
Twenty	Stop	Twenty
Stop	Stop	Stop

State Transition Diagram

A **state transition diagram** is a graphical representation equivalent to the transition matrix. The state transition graph uses two elements: a circle to denote the state and an arc for the transition. (Technically, for mathematicians, the circles represent vertices and the lines represent edges of a directed graph.) In a very simple diagram the arcs might be straight lines. The transition condition is written over the arc. The state transition diagram for the vending machine counter is shown in Figure 4.3.

A transition matrix is a table and therefore in some situations it is easier to draw than a graph. On the other hand, a state transition diagram is more readily comprehensible than a transition matrix. In case of a Parser both presentations show the same information and are 100% equivalent. The main task of the state machine as discussed in this book is to generate actions. Later we will see that unfortunately neither a transition matrix nor a state transition diagram can show all design details of a state machine with actions. In such a case we need another presentation means.

Note that we have bypassed a full discussion about the state *Twenty*, just mentioning that the coin 10 is ignored there. A complete design requires some actions to be done (reject the coin or give change) in this case. For simplicity we ignore these details.

Actually, the solution for the Vending machine counter example does not have too much practical value. It would be a typical specification for a coded solution where we assume that during the coding we find a proper implementation. The transitions are defined assuming that the information about each coin (5, 10) is a true event; i.e., it triggers the state machine execution environment and then disappears. Otherwise, a single value 5 applied in the state *Start* would cause the transition to the state *Five* where the transition condition would be immediately fulfilled and the state machine

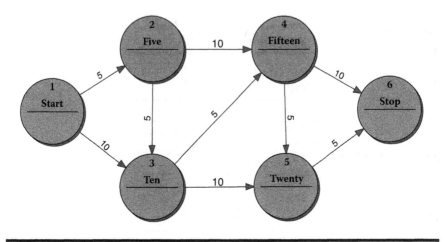

Figure 4.3   The state machine diagram for the vending machine counter.

would go to the state *Ten*, from there it will go to the state *Fifteen*, and so on until it reaches the state *Stop*. A specification must therefore be matched to the state machine execution model that will be used (see also the discussion in the section *State machine execution models* in Chapter 6).

Outputs (Actions)

Applications that can be served by a Parser model are not many. First, the idea of an initial state *Start* and a terminating state *Stop* is very limiting — a state machine has to work continuously, all the time, in the general case. The other important aspect: most applications require some actions to be performed (outputs); actions required by the controlled system. Note that in hardware design the term "output" is commonly used, while for software design the term "action" is more popular. Several types of actions can be defined depending on the conditions and moment they are performed:

■ Entry Action
■ Exit Action
■ Input Action
■ Transition Action

The **Entry Action** is an action done when the state machine enters a state. The **Exit Action** is an action done when the state machine leaves the state. The **Input Action** is an action done when an input (condition) is true. Each state has its own set of Input Actions. Input Actions that are done in any state (effectively state independent) are also used. The

Transition Action is an action performed during the state change. Note that although similar to them it is neither an Entry Action nor an Exit Action, which are both state dependent; the Transition Action is transition dependent.

We use various Actions to make a state machine design understandable, and we apply certain rules. If a state machine changes state, all actions: Input, Exit, Transition, and Entry Actions are carried out in this sequence but practically in the same moment. Without a state change only an Input Action may be performed.

Moore and Mealy Model

In practical situations not all these actions are used. Depending on the actions which are used some typical models have been defined; the best known are the **Moore** and **Mealy** models.

A state machine that generates only Entry Actions is called a Moore model. A state machine that generates only Input Actions is called a Mealy model. The models selected will influence a design, but there are no general indications regarding which model is better. Choice of a model depends on the application, execution means (e.g., hardware systems are usually best realized as Moore models), and personal preferences of a designer or programmer. In practice, mixed models are often used with several action types.

State Transition Table

The existence of Actions complicates the presentation problem. Using a transition matrix or state transition diagram it is difficult to express the functionality of a state machine with several actions. Therefore, we introduce a **state transition table** as the most versatile tool for expressing the complete specification of a state machine.

We use the state transition table shown in Figure 4.4. The table contains fields to specify Entry, Exit, and Input Actions. To keep the table as simple as possible there is no explicit field for Transition Actions — an Input Action with its condition equal to a transition condition could be treated as a Transition Action. A more detailed discussion would show that this arrangement does not correspond fully to a Transition Action definition. We return to this later.

Each State has its state transition table. The table consists of several fields used to specify actions and transitions. Each Action field may contain several Actions. A table may contain several **Input Action expressions**

State	Entry action	Entry_Action1 Entry_Action2
	eXit action	Exit_Action1 Exit_Action2
	Input_Action_Condition1	Input_Action1 Input_Action2
	Input_Action_Condition2	Input_Action3
Next_State1	Transition_Condition1	
Next_State2	Transition_Condition2	

Figure 4.4   The state transition table for one state.

consisting of related Input_Action_Condition and Input_Action fields. Similarly, a table may contain several **Transition expressions** consisting of several Next_State and Transition_Condition fields. In the condition fields we will use two Boolean operators: AND (represented by &) and OR (represented by |) to define more complex logical expressions. The usage is as intuitively understood:

`this_control_value & that_control_value` or

`this_control_value | that_control_value.`

If the state transition table is to express completely the behavior it should also have well defined priority rules for Transitions and Input Actions.

> The rule for transitions is obvious* as it is impossible to make more than one transition at the same time we agree that the sequence in the table determines the priority.

For instance, if both: `Transition_Conditions1` and `Transition_Conditions2` are true the transition to Next_State1 will be performed.

The priority rule for the Input Actions is not so obvious and depends on the execution environment.

> If we assume that all Input actions that are due will be performed, then their sequence in the table must not play any role.

---

* We are not interested in a *nondeterministic* model mentioned later in the *Models of a finite state machine* in Chapter 6.

Table 4.3   Transition Matrix with Input and Transition Actions

To *From*	...	*State1*	*State2*	...
...				
`State1`		condition1/Input Action1	condition2/Input Action2	
...				

Otherwise we would try to realize a control sequence by prioritizing Input Actions: it must be done by adding a state or two. This is the rule we follow in this book (the priority would be important if the execution environment performs only one Input Action at any one time).

For comparison let's draw a transition matrix and a state transition diagram with Actions. Considering a transition matrix there is no agreed convention to show all actions, so only Input and Transition Actions can be specified there as shown in Table 4.3.

The content of a transition matrix specifies both the condition and the action. If the cell coordinates are the same (e.g., *State1* in the example) the cell specifies the Input Actions (the state does not change). If the cell coordinates are different (e.g., from *State1* to *State2* in the example) the cell specifies the transition and the Transition Action.

It would be theoretically possible to show all Actions in a state transition diagram, but due to practical sheet size it would not make much sense. Occasionally an expression like

Transition Condition/Transition Action

is used to display Transition Actions.

We may now return to the equivalence of Transition Action and Input Action. Let us assume that we have a *State* where the `Condition2` for an Input Action equals a transition `Condition2` and there are two transitions possible as in Figure 4.5. If only the `Condition2` is true the Action will be carried out and the state machine also changes to *State2*. In that case we could call the Action a Transition Action bound to the transition from *State* to *State2*. If both conditions: `Condition1` and `Condition2` are true, the Action will be performed of course but the transition to *State1* will be done. In that case we have to regard Action as a Transition Action bound to the transition from *State* to *State1*. Of course, it does not make sense. This example shows that the equivalence between Input Action and Transition Action is rather superficial.

The diagrams in Figures 4.5 and 4.6 show states and transition conditions for which the state transition table of the state *State* would correspond

State	Entry action	
	eXit action	
	Condition2	Action
State1	Condition1	
State2	Condition2	

Figure 4.5   An Input Action is not a Transition Action.

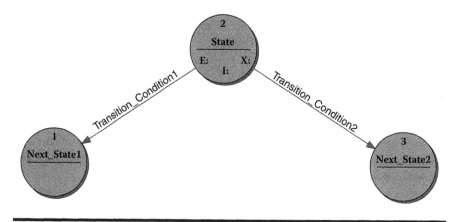

Figure 4.6   The state transition diagram.

to that in Figure 4.4. The existence of any actions are indicated by the letters E: for Entry, X: for Exit, and I: for Input Actions. A missing letter means that there are no actions of that type.

It is obvious that a state transition diagram gives an overall impression about the state machine, displaying the states and transitions quite well but the details are hidden, especially those of the Actions. Thus, a state transition diagram needs to be supported by state transition tables that contain the full specification of states.

Usage of a transition matrix is limited to Mealy models. It would be possible to introduce a convention of writing Entry and Exit Actions in the transition matrix in the column (*To*) labels but it has never been done. The reason is obvious: the complexity and the size of the matrix will become too large as it effectively would imply the transition matrix contains the content of all state transition tables. It would work for some textbook examples but it will be no use for any practical design.

This review of three presentation tools shows that a combination of a state transition diagram and state transition tables is a reasonable compromise between the desire to have a comprehensible picture of state machine functioning and the requirements to have a complete specification.

Example

We show in Figures 4.7 through 4.13 an example without trying to explain the requirements. We just want to show the presentation means used in this book. The presentation starts with a state transition diagram that gives an overall idea about the state machine: its states and transitions. The specification details are contained in the following state transition tables; there is a separate table for each state. The first table *Always* specifies the state-independent Input Actions performed at any time if due. State tables begin with (optional) comment fields.

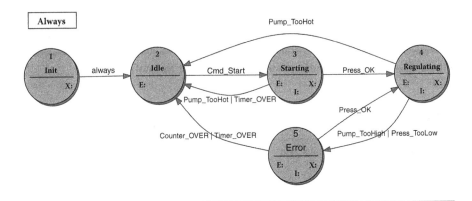

Figure 4.7 Example of a state transition diagram.

Specification of state independent actions. If the Required Pressure value changes:
- the Error Counter is reset,
- the Pressure Limits are recalculated,
- the Pressure value is set.

Always	RequiredPress_CHANGED	Counter_ResetStart Ofun_CalcLimit SetPressure_Set

Figure 4.8 Example: the state transition table of the state *Always*.

Init state is the initial state. Once left it will never be reached again.

Init	Entry action	
	eXit action	Swip_On
Idle	always	

Figure 4.9   Example: the state transition table of the state *Init*.

All activities are ceased. Waiting for a Start command.

Idle	Entry action	SetPressure_Off
	eXit action	
Starting	Cmd_Start	

Figure 4.10   Example: the state transition table of the state *Idle*.

Several activities are initiated. Waiting for Pressure acknowledgements. Due to a Timer missing acknowledgement leads to return to the Idle state.

Starting	Entry action	MyCmd_Clear SetPressure_Set Counter_ResetStart Timer_ResetStart Ofun_CalcLimit
	eXit action	Timer_Stop
	RequiredPress_CHANGED	Timer_ResetStart
	Pump_TooHot	Al_PumpTooHot
	Timer_OVER	Al_PressureError
Regulating	Press_OK	
Idle	Pump_TooHot | Timer_OVER	

Figure 4.11   Example: the state transition table of the state *Starting*.

The Pressure value is ok.

Regulating	Entry action	LED_On
	eXit action	LED_Off
	Pump_TooHot	Al_PumpTooHot
Error	Press_TooHigh | Press_TooLow	
Idle	Pump_TooHot	

Figure 4.12   Example: the state transition table of the state *Regulating*.

The Pressure value is outside limits. A return to the state Regulating is possible if the Pressure improves in a certain time. A Counter forces a return to the Idle state if the number of errors exceeds a predefined value (see project settings). A Timer ensures that a missing acknowledgement will cause a transition to the Idle state.

Error	Entry action	Timer_ResetStart
	eXit action	Timer_Stop
	RequiredPress_CHANGED	Timer_ResetStart
	Counter_OVER | Timer_OVER	Al_PressureError
Regulating	Press_OK	
Idle	Counter_OVER | Timer_OVER	

Figure 4.13   Example: the state transition table of the state *Error*.

Recommended Reading

1. Gomez, M., "Embedded systems programming feature," *Embedded Systems* 13, no. 13 (December 2000).
2. http://en.wikipedia.org/wiki/state_machine.

Chapter 5

Hardware Applications

Introduction

The book is about state machines in software. But state machines were known and used before software emerged. Especially, designers of digital hardware systems used, from the beginning, Automata Theory methods in their work; among others, they used state machines for constructing sequential circuits. Considering the closeness of hardware and software we have decided to insert a few pages that should give a reader at least an impression of hardware design based on state machines. The complete knowledge about the use of state machines in designing digital systems can be found in several publications; the books of Commer[1] and Sunggu[2] are examples of useful references.

Limited to Boolean Signals

Hardware systems are built using gates and flip-flops, i.e., digital components that operate on two-valued signals: true and false. The functioning of digital systems is described by rules of the Boolean algebra. Typical examples of digital components are shown in Figure 5.1.

The gates and flip-flops are mostly packed into larger digital components: EPROM, FPLA, etc.

The design of digital hardware systems is based on Automata Theory methods which, if properly applied, result in optimal systems. The definition of optimal systems depends on the requirements, where the important

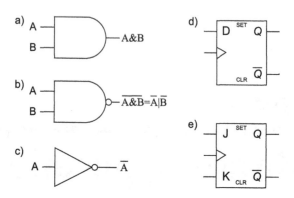

Figure 5.1   Digital components: (a) AND gate; (b) NAND gate; (c) NOT gate; (d) D flip-flop; (e) J-K flip-flop.

factors are speed and number of used components. Although the requirement to build minimal systems decreases with the enormous increase of the hardware features, it still makes sense to build optimal hardware systems.

We are interested in state machines in software. So, we do not discuss the methods used in detail. We limit ourselves to showing an example of a hardware system design to get some feeling for differences between the software and hardware approaches.

A digital system can be asynchronous or synchronous. Asynchronous systems are fast and can be realized with a minimal number of components but they are difficult to design, test, and suffer from timing problems (hazards), which are a true challenge for designers and difficult to get under control.

Synchronous systems have a clock that dictates the speed. They are slower than asynchronous ones but essentially easier to design, and therefore they dominate the world of digital circuits. The example will be designed as a synchronous system.

Design Example — Traffic Light Control

We want to design a traffic light control at a level crossing of a railway and a road. As the details of the requirements may differ slightly we define them as follows: There is one rail only but the trains may come from both directions. The trains are detected by 3 sensors L (left), M (middle), and R (right) positioned as in Figure 5.2. We assume also that the distance between two sensors is larger than the longest train, in other words a train can never activate 2 sensors at the same time. The output Y of the system is a red lamp that should be switched on if the train approaches the sensor L if coming from left and should be switched off if the train

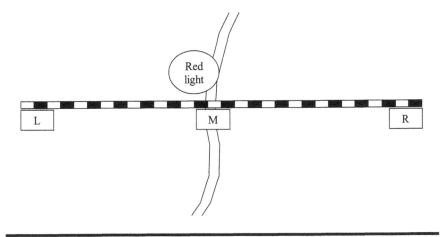

Figure 5.2   The traffic light control at a level-crossing of a railway and a road.

leaves the sensor M. Similar actions should be performed if the train comes from the right.

After some trials, we found the state transition diagram shown in Figure 5.3, which describes adequately the control of the traffic lamp. The design is simplified and does not cover all possible error situations — it takes into consideration the *sunny-day-scenarios*. Typically for the hardware design we use the Moore model for the control where the state will determine the output Y.

In the diagram the signal X in the transition conditions means: X = L | R.

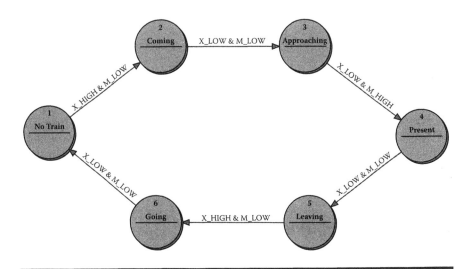

Figure 5.3   The state transition diagram of the traffic light.

Table 5.1 Coding of States for the Traffic Light Control

Description	State	$Q_2Q_1Q_0$	$D_2D_1D_0$ $XM =$ 00	01	11	10	Y
No train	NoTrain	000	000	—	—	001	0
On X	Coming	001	010	—	—	001	1
Between X and M	Approaching	010	010	011	—	—	1
On M	Present	011	100	011	—	—	1
Between M and X	Leaving	100	100	—	—	101	0
On X	Going	101	000	—	—	101	0

where: D_n corresponds to Q_n^+ which is the next state of Q_n

Coding of the 6 states requires 3 flip-flops. If we choose, e.g., the coding shown in Table 5.1, we get the following logical equations for implementation with D flip-flops (the corresponding Karnaugh tables are omitted — see Appendix F: *Traffic Light Control — Design of the Hardware Solution*):

$$D_2 = Q_1 \& Q_0 \& \overline{M} \mid Q_2 \& X \mid Q_2 \& \overline{Q_0}$$

$$D_1 = \overline{Q_2} \& \overline{Q_1} \& Q_0 \& \overline{X} \mid Q_1 \& \overline{Q_0} \mid M$$

$$D_0 = X \mid M$$

$$Y = Q_1 \mid \overline{Q_2} \& Q_0$$

In the equations, as in the entire book, we use the symbol *&* for the Boolean AND operator and the symbol | for the Boolean OR operator.

The system can be realized using NAND gates and D flip-flops as shown in Figure 5.4. To keep the schematic simple we omitted from the schematic the lines from the flip-flops to gates. Nowadays, we would rather implement the above system using programmable read-only memories (EPROM), complex programmable logic devices (CPLD), or field programmable logic arrays (FPLA).

EPROM-Based Implementations

EPROM-based implementations use a register of D-flip-flops; the transition conditions and output function are coded into a read-only memory. Figure 5.5 presents an EPROM-based solution of the discussed example.

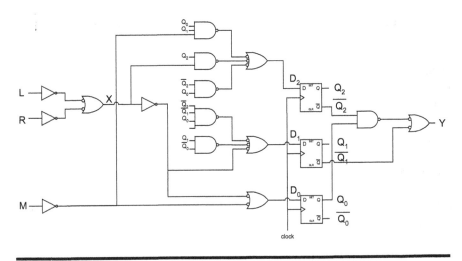

Figure 5.4   The traffic light control realized with D flip-flops and NAND gates.

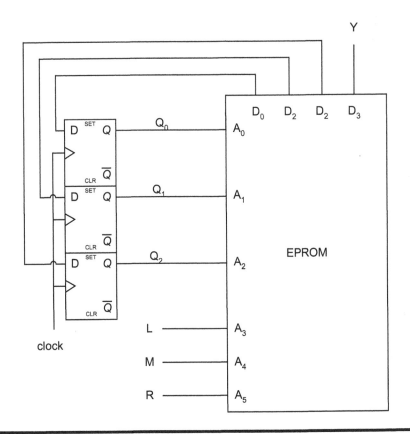

Figure 5.5   EPROM-based implementation of the traffic light example.

The states of the state machine are stored in a 3-bit register, which corresponds to the D flip-flops. The outputs of the register Q_0, Q_1, Q_2 plus the sensor inputs: L, M, R create the address of the EPROM. The EPROM data outputs represent the next state ($D_0D_1D_2$ -> $Q_0^*Q_1^*Q_2^*$) and the output Y.

Alternatively, we could do without coding of states and use 6 bit-register where each address bit corresponds to one state according to the state transition diagram in Figure 5.3. This so-called 1-hot design is simpler, and for small state machines makes sense because EPROMs and register are today very cheap. This approach can be applied to rather small state machines — above a certain number of states (and inputs) coding of states is required.

The content of the EPROM is shown in Table 5.2. The table shows only the content for relevant addresses; other addresses are never used and may contain default value 0 or 1.

Table 5.2 The Content of the EPROM in Figure 5.5

A_5 (R)	A_4 (M)	A_3 (L)	A_2 (Q_2)	A_1 (Q_1)	A_0 (Q_0)	D_3 (Y)	D_2 (Q_2^*)	D_1 (Q_1^*)	D_0 (Q_0^*)
0	0	0	0	0	0	0	0	0	0
0	0	0	0	0	1	1	0	1	0
0	0	0	0	1	1	1	1	1	0
0	0	0	0	1	0	1	0	1	0
0	0	0	1	0	1	0	0	0	0
0	0	0	1	0	0	0	1	0	0
0	0	1	0	1	1	1	0	1	1
0	0	1	0	1	0	1	0	1	1
0	1	0	0	0	0	0	0	0	1
0	1	0	0	0	1	1	0	0	1
0	1	0	1	0	1	0	1	0	1
0	1	0	1	0	1	0	1	0	1
1	0	0	0	0	0	0	0	0	1
1	0	0	0	0	1	1	0	0	1
1	0	0	1	0	1	0	1	0	1
1	0	0	1	0	1	0	1	0	1

FPLA- or CPLD-Based Implementations

CPLDs contain a number of "logic blocks," which each have a few flip-flops and input logic that has a number of AND–OR units, able to generate quite complex expressions from 20 or more inputs. Thus, an implementation with programmable logic corresponds to the implementation with flip-flops and gates. In such a case, we do not build the system using discrete components but we use specialized programs to specify the system and to program the array. Similar considerations apply to FPLAs, which typically have a less structured basic design in silicon, although a similar structure would be imposed by the design software tools.

Conclusions

A state machine concept stays the same, independently of the application domain area. The differences are in the development culture resulting from the implementation features. Errors in hardware design are expensive. A hardware designer cannot "try" too often as the result of the designer's work is transformed into a physical device that can be effectively tested. Only a limited number of design iterations make economic sense. Thus, hardware designers must use methods that increase the probability of correct design.

Contrary to this, software developers assume that they are allowed an infinite number of iterations. Hence, the pressure to strive for a first-time functioning solution is lower.

The comparison is probably not quite fair. Complex hardware systems are also difficult to design on paper. Their functioning must be tested using simulators before they go into production. Testing means nothing else but trying to find a proper solution, which we could not find in a more methodological way. In any case, the cut between design and manufacturing in hardware is very clear: only a designed system goes into production. In software the difference between design and manufacturing is rather vague, which very often leads to chaotic development.

Hardware systems are essentially less complex than software applications. In spite of impressive hardware achievement, the systems are not very sophisticated. They are large and very reliable but the functional sophistication is made only by software. Therefore, we would not try to make a system as in the TrafficLight example more complex covering all imaginable (error, unexpected) situations. The transition from the specification to the implementation is just too difficult — it would blow up any design method very soon. Using software solutions we are able to cover much more complex situations. The example presented in this chapter is really trivial for software. If we compare the example from this chapter with a similar example shown in *Example — Traffic light control* in Chapter 9, we see that in software we may realize much more difficult problems. The irony is that we believe sometimes that we may do it without any method, just using our common sense.

In practice the above example would no doubt be implemented nowadays in software, using a micro-controller, and hardware finite state machines are more commonly found within integrated circuits.

Recommended Reading

1. Commer, D. J., *Digital Logic and State Machine Design*. New York: Oxford University Press, 1995.
2. Sunggu, L., *Advanced Digital Logic Design Using VHDL, State Machines, and Synthesis for FPGA's*. Florence, KY: Thomson-Engineering, 2005.

Chapter 6

Software Specific

Introduction

If we take two books that have in their title "Automata Theory," but one book is written for hardware designers and the other one for software people, we get the impression that there are two different Automata theories. The hardware book is about digital system design. The software book is about mathematical description of computations, especially programming languages. Both use the concept of finite automaton, also named finite state machine (or the short form state machine we use in the book), as the basis of their methods and theorems but the terminology and the emphases are completely different:

- In hardware design a state machine is a vehicle for a design method.
- In software a finite automaton is a means to prove theorems.

To illustrate this point let us take the concept of deterministic and nondeterministic automata. From the hardware point of view that differentiation is irrelevant because only deterministic automata are used in design methods. For computer language discussions of a nondeterministic nature automaton is a useful concept used for proving theorems.

Both hardware and software solutions for control problems require the same scientific foundation. Specifying application behavior with the help of state machines we should expect to use exactly the same means: there is no reason to have two different specifications for hardware and software implementation. In other words, both hardware and software can use the same finite automaton concept.

The book is about the use of state machines in software. We present here a picture of a state machine that is needed for specification of behavior. But we have not written another book about Automata Theory. This statement explains the structure and content of that chapter. We discuss here various aspects of state machine definitions and features trying to show which of them are relevant for our aim: the use of state machine for specification. The software specifics play a role in that discussion, of course, and will be the starting point in the following sections. But the software specifics refer to the specification environment and not to a model of a state machine. Models of state machines in the book are not hardware or software specific as there is only one concept of a state machine independent of its use.

Data and Control Flow

Hardware logic systems operate in principle on purely digital inputs and outputs. Software systems are more versatile — they process any sort of data. Therefore, the application of the state machine concept to software design is not so obvious and easy.

A state machine is a decision machine; it can react only to input stimuli (expressions) that carry clear information: true or false.

> A common approach in application design is to use the state machine model as an auxiliary tool to specify a local, limited task. This specification is done rather superficially knowing that the implementation will be coded anyway where all details needed for integration into the entire software will be handled. In other words, taking an application as a whole, a designer has to analyze simultaneously both application- and software implementation-dependent details.

For example: analyzing how to switch a motor on or off and to guard it against overheating a designer has to solve such problems as:

- How to realize and handle the time-out
- How to get the motor temperature information and to filter out the "overheating" signal
- How to signal a problem to the operator, etc.

The result of this approach is a confusion that makes it difficult to distinguish between logical (application control relevant) errors and software coding errors.

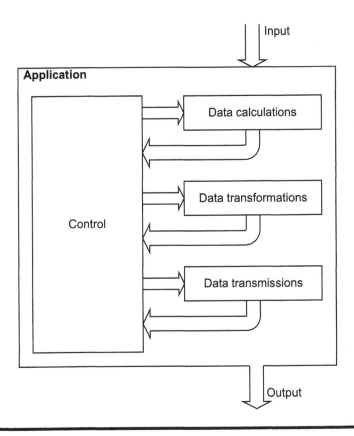

Figure 6.1   The data and control flow in application design.

The first step in more effective use of state machines in an application design is to comprehend and accept the partition between the data flow and control flow. Application software seen from this perspective would present a picture as in Figure 6.1.

The Control represents the **control flow**. The **data flow** is created by Data processing blocks, which are passive: they cannot do anything on their own but are triggered by signals generated by the Control to perform calculations, transformations, transmissions (these are just examples of operations that may be done on data). Results of data processing are inputs used by the Control.

This picture may seem incorrect to an object-oriented programming (OOP) purist but we do not speak about the software implementation at this moment. The software implementation may indeed be the product of a wonderful OOP design process. This presentation means that on the application specification level we should think first of all about the application and construct a behavioral model which separates the Control from

Data processing. We may apply this approach to a pure software design which will be discussed later.

Any Class of Signal May "Contain" the Control Value

Let's take some typical data (signals) and show their possible control content — we shall filter out the control feature and call it a **control value**. The control value "describes" a data property that can be used for control purposes. A control value is not the same as a Boolean variable. One item of data may be assigned several control values that are typically mutually exclusive but not necessarily so — sometimes more than one may exist at a given moment.

Digital Input

A digital input is a signal that has two values corresponding to Boolean terms `false` and `true`. A digital input has two control values; let's call them `Off` and `On`, which correspond to `false` and `true` values. The `Off` and `On` control values are similar to Boolean `false` and `true` values but are not strict equivalents. The Boolean values are values of a variable. The control values `Off` and `On` are features of data (in that case digital input). It is true that the `Off` value excludes the existence of the `On` value, and vice versa. On the other hand, it is imaginable that neither value `On` nor `Off` is asserted — this situation arises if we have no knowledge about the digital input.

The definition of a control value may seem a bit unusual, when considering Boolean algebra. This approach has a significant advantage — by adding another control value `Unknown` we get a fuller description of a digital input, e.g.:

```
Di_Off, Di_On, Di_Unknown.
```

The conventional approach of considering a digital input as a simple Boolean variable fails in undefined situations: if we do not know the value of a digital input we have to assume the value, setting it arbitrarily to `true` or `false`. For more complex data types the advantage of the control value idea will become more obvious: it is the only definition having no Boolean equivalents.

Command Input

A command input is a signal that has several values, e.g., names or integers that mean "do this" or "do that." It is obvious that it is difficult to speak about the direct use of command values in Boolean expressions; we can do it only indirectly, e.g. (using C syntax):

```
If (command == Cmd_Start) then ...
```

or we might prefer a switch statement.

Using the concept of a control value we say that a command has, e.g., the following control values:

```
Cmd_Start, Cmd_Stop, Cmd_Continue, Cmd_Unknown.
```

Instead of operating on a single command value we use in this example four control values. In this case they are mutually exclusive as the existence of two commands at the same time would be rather difficult to interpret.

Numerical Input

A numerical input is a signal that represents a physical quantity such as temperature, voltage, pressure, or some other so-called analog quantity. Typically, a numerical input is an integer or floating point number and as such there is no way to treat it directly as a Boolean value, especially as its range is essentially infinite. We overcome this difficulty when coding, as for instance:

```
If (value > Limit_High) then ...
```

Using the concept of a control value we can say that a numerical value might have, e.g., the following control values:

```
Temp_Low, Temp_High, Temp_Ok, Temp_Bad, Temp_Unknown.
```

Instead of operating on a single numerical value we use in this example five control values. In this case they are not mutually exclusive. It is imaginable that both values, `Temp_Low` and `Temp_Bad`, do exist at a given time. The exact interpretation of the control values is outside the discussed control scope: the meaning is defined by a definition of some limits. It is a nice example showing what we understand as separation of the control flow from the rest of the system. For the control flow, only control-relevant issues (the temperature range: for instance `Temp_Low`) make sense but not the absolute level of the temperature. The low limit defining the control value `Temp_Low` might be 7.8V corresponding to 5°C in one case and a completely different value in another case but those fine details, although obviously important, are completely irrelevant for the control system design: they are just properties.

Parameter

A parameter is an item of auxiliary data specific to a control process or system. It can be of any type: Boolean, integer, float, etc. Normally, the value of a parameter has no control meaning. On the other hand, a change of its value, its usage, or initialization may be relevant for control. In such a case we can define, e.g., the following control values for a parameter:

`Par_Init, Par_Changed, Par_Used.`

In most cases the use of a `Par_Unknown` control value would not be justified — parameters are typically included in software; they are not inputs. If parameters are delivered from the controlled environment and are not accessible at a system start we could use the further control value `Par_Unknown.`

Data Processing Result

A data processing result indicates whether a data calculation, transformation, or transmission has been a success or a failure. Typically, a data processing result is an integer. We can define, e.g., the following control values for a data processing result:

`Result_Success, Result_Failure1, Result_Failure2,`
`Result_Failure, Result_Unknown.`

The control value `Result_Unknown` does make sense and is the initial value before the Data processing is performed. A name like `Data_NotProcessed_yet` might be an alternative choice.

Timer

A timer is an object counting time and signaling the expiration of a predefined period (time-out). A timer belongs to a larger class of objects — counters. Counters count events; a timer counts specific events: time clocks.

When a timer expires it generates the `Over` (or `overflow`) signal. It is the most obvious signal used in control. In general, other timer states (i.e., control values) may be of interest. For instance, we can define the following control values for a timer:

`Timer_Over, Timer_Running, Timer_Stopped,`
`Timer_NotAvailable.`

Because a timer belongs to computer resources, its state must be always known — therefore the `Timer_Unknown` control value is superfluous here. If the software cannot find out the state of the timer it is corrupted or has errors and must be debugged. Note the difference between the `Timer_NotAvailable` and `Timer_Unknown` control values: the first can be generated by a system start-up only and means that the computer resources are exhausted; the second would be just a software bug.

From the control point of view, the character of the signals (the `Over` signal is typically an event, the `Running` or `Stopped` values must be requested) does not play any role. It is another point in the discussion about data and control flow separation. We want to filter out only the control relevant feature for control specification. We want to "forget" all factors that depend on software implementation, operating system, development environment, etc. The control flow must be governed by pure control information.

State Machine

In the book we use several signal types and discuss their control values. Especially, in Part III we have a very complete overview of RTDB objects and their control values. We end the examples in this chapter by mentioning state machines. The states of state machines are by definition control values. For instance, we can define the following control values for a state machine Main using states of its Slaves: Device, Pressure1, and Pressure2 (see *Example — Pumps supervision system* in Chapter 9 or Appendix L *Pumps supervision project*):

```
Pressure1_Error, Pressure1_Idle, Pressure2_Error,
Pressure2_Idle, Device_Start.
```

Any state machine may use states of other state machines exactly as digital inputs, timers, etc. to define its Transition or Input Action conditions.

External and Internal Signals

Control values represent features of input signals. The source of inputs may be internal or external in relation to the control software. Digital inputs or numerical inputs are by definition external signals. Timers are good examples of generators of internal signals. Both categories of input signals generate control values that differ in one aspect: the external signals may be unknown. In contrast to external signals, the control value of internal signals must always be known; otherwise we have a software bug. This difference influences the design: the links to external signals must be continuously supervised to see whether they stay intact and supply the signal changes; in case of interruptions the value of external signals must get the value Unknown.

What about Outputs?

Up to now, we have discussed Control inputs as they are used to define the behavior. Their counterparts — outputs — are produced by the

Control. Outputs in the control flow are descriptions (names) of activities. In contrast to inputs, the outputs are not used for control but are results of commands which we call actions.

Digital Output

A digital output is a signal transferring two values corresponding to the Boolean `false` and `true` to the controlled application. The Control has to generate two actions: `Off` and `On`. These actions set the two values of the digital output. The digital output must have a defined value at a system start — it must be set arbitrarily to `false` or `true` value. Once set it keeps the value until reset, and vice versa. Thus, the actions of a digital output can be considered triggers that change the output; if actions are absent, the output keeps the last set value. Typically, the digital output actions are as follows:

`Do_Off, Do_On.`

Command Output

A command output is a signal generated by the Control and used to pass control information to other Controls (state machines). Note that a command output produced by one state machine is typically a command input for another state machine. Examples of command outputs may be actions:

`Cmd_Start, Cmd_Stop, Cmd_Continue.`

Numerical Output

A numerical output is a signal (action) responsible for setting physical quantities: temperature, voltage, pressure, etc. These quantities are typically integers or floating-point numbers and therefore they cannot be set directly by the Control. The Control can decide only when to set a numerical output, as the numerical details are arranged outside the scope of the Control. Hence, numerical outputs are, e.g., actions (commands to set):

`Pressure_Off, Pressure_Set.`

At a system start a numerical output must be initialized to a defined value, typically to the 0 value.

Data Processing Output

A data processing output means a call of a function that has to be triggered by an action. The data processing output (action) may have different flavors, as in the following example:

`Calculate_with_parameter1`, `Calculate_with_parameter2`,
`Calculate_Initialize`.

Here, the actions are commands to do some calculations.

Timer

A timer has already been used as an example in the discussion of inputs. The timer contains both inputs and outputs: it must be triggered by actions (e.g., to make it run) and produces control values (e.g., the `Over`). The timer may be triggered by the following actions (commands to do something):

`Timer_Start`, `Timer_Stop`, `Timer_Reset`.

At a system start a timer must be initialized to a certain state; typically a timer is reset and ready for a `Timer_Start` command.

Summary

A state machine, as it is a decision system, can operate on Boolean values only. Hence, only digital values are directly suitable for its control. The concept of a control value allows us to treat any control relevant data in a way similar to Boolean inputs. The above-discussed data types are just examples. This approach can be applied to any data that may influence the control and is to be used by a control specification.

Both involved terms, control inputs and actions, represent pure control features. Inputs are represented by control values which contain only control relevant information. Outputs are represented by actions, which are orders of what to do without any implementation details. This arrangement allows a software control system to be treated in a way similar to a hardware control system. Hardware state machines can be designed using Boolean algebra due to their pure Boolean input and output signals. Analog values in such systems need to be processed, e.g., by a comparator, to generate the digital value. Software control systems are more demanding as their inputs and outputs are more varied. In any case,

> with the introduction of control values for Inputs and actions for Outputs we have created a basis for a similar (to hardware) approach in the software world.

The strict partition between the data and the control flow is the basis for a completely different approach to the design of software-based control system applications. Without such partitioning, software designers consider state machines and similar concepts a way to specify some pieces of software, but they do not treat them as means to build the control flow of the entire software. This is a severe underestimation of the power of the state machine concept. In contradiction to commonly accepted ideas on this topic, for large and complex projects a system of state machines may be used to determine the behavior of the entire program, contributing essentially to higher software quality.

Event-Driven Software

Software "interfaces" with the controlled application exchanging data: supplying input data and receiving output data. There are two basic interface mechanisms used: polling and event driven.

Polling means that inputs are cyclically checked and outputs are cyclically set. This arrangement is predictable and guarantees the reaction time, which is defined by the polling cycle.

Event driven means that the interface generates events if inputs change; outputs are also set if their values change. Event-driven systems are favored in software as a more effective (software does something if necessary) and on average faster solution, although the event-driven system cannot guarantee any response time. Events are also generated internally by computer resources, e.g., an expiring timer generates an Over event. From the control point of view there is no difference between external and internal events.

An event is a trigger that says that something has changed — the trigger should be used to get the control information from the changed input. Sometimes the event itself carries the control information.

Event as a Control Signal

The programmers' love of events produces some strange effects. Several software implementations based on the state machine concept assume that events are the only signals that can be used to control state machines, and implicitly that events that are not consumed (i.e., do not immediately provoke an action or transition) are discarded. That assumption means that states of a state machine must "store" not only the history of input changes (in a very compressed form) but the present value of inputs as well. In other words, the state transition diagram must explicitly contain all possible control paths. We discuss the problem in Chapter 7 *Misunderstandings about FSM* in the context of the so-called "state explosion" phenomenon; therefore, we limit the discussion at this moment to a simple example of an outdoor security lamp.

This should light when movement is detected, and remain lit for, say, 10 seconds after movement ceases. But it should not light in daylight, so

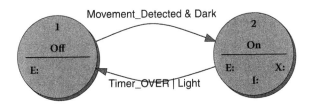

Figure 6.2   Security lamp: the state transition diagram.

Off	Entry action	Lamp_Off
	eXit action	
On	Movement_Detected & Dark	

Figure 6.3   Security lamp: the state transition table of the state *Off*.

On	Entry action	Lamp_On Timer_ResetStart
	eXit action	Timer_Stop
	Movement_Detected	Timer_ResetStart
Off	Timer_OVER | Light	

Figure 6.4   Security lamp: the state transition table of the state *On*.

it incorporates a sensor of the ambient light level. Using a classical fsm model, we need just two states, the initial state *Off* and the state *On* (Figure 6.2 through Figure 6.4).

The lamp controller passes to *On* when movement is detected, but only when the ambient light level is low. An Entry Action in the state *On* switches on the lamp and starts a timer, and an Input Action re-starts the timer whenever movement is detected. In the state *On* the transition to the state *Off* occurs when the timer reaches the time-out delay, and perhaps also when the ambient light level has increased. The Entry Action in the state *Off* switches off the lamp.

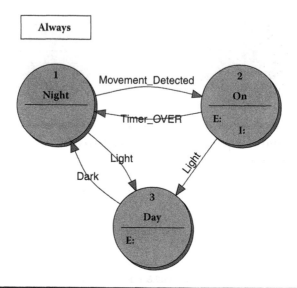

Figure 6.5   Security lamp as a pure event-driven state machine: the state transition diagram.

Now consider the equivalent state transition diagram for a purely event-driven version (Figure 6.5). It requires an additional state *Day* to store the information about the day. The state *Night* corresponds to the state *Off* in the first solution. On entry to either the state *Day* or *Night* the lamp is switched off. The state *On* is equal to the state *On* in the first solution.

Overloading the state machine with the burden of storing the value of present inputs cannot work except for some trivial examples like the Security Lamp control. Read more about it in Chapter 7 *Misunderstandings about FSM.*

State Machine or Combinational System?

We use the example of the Security lamp above to discuss the problem of the size of a state machine. To be more specific — we should like to find the answer to the question: how many states should a state machine have?

The way we have started to specify the example is very typical — we have just assumed that the behavior of the discussed control problem can be described by a state machine and we "see" immediately the obvious states, in that case: *Off* and *On*. Very seldom we ask ourselves the question whether it could be realized as a combinational system. In a way, it is understandable as only very trivial behavior has a combinational character and, eventually, who cares about the number of states in software? The other factor that influences the specification is the implementation — a behavioral model must be consistent with the execution system. If we intend to code the specification by hand, we tend to make a rather superficial state machine design, as we know that the "hour of truth" will come in the coding and we treat the specification as a useful introduction and help, but not as the ultimate solution. If the specification is to be executable using a ready execution environment, we cannot neglect its features — the state machine must fulfill exactly all the requirements of the runtime system.

We can now present another solution for the Security lamp, shown in Figure 6.6. Two Input Action expressions are required to control the lamp. The first expression says that if the movement is detected by night, the lamp is switched on and a timer restarted. The second expression says that if the timer expires or it is not dark, the lamp is switched off and the timer stopped. This solution does not require any states to store input changes — the present control values of daylight (dark or light), move-ment, and timer contain full and clear information about the inputs. But we have to admit that to "see" that it can be a combinational system is not so obvious and that the solution is less clear than the previous ones. As a rule, more compact solutions are a bit tricky and are found by a more experienced designer.

To be quite correct we should also note that because of the timer the system is still a sequential circuit — only the timer contains the sequential part and in fact it is a state machine of some sort. Discussing the StateWORKS implementation we will see later that using powerful objects (that are themselves state machines) simplifies the specified system.

We presented three different solutions of the Security lamp control, but with exactly the same behavior: with 1 state (combinational system — Figure 6.6), with 2 (Figure 6.2), and with 3 (Figure 6.5) states. Those solutions demonstrate the relativity of so-called optimal or minimal solu-tions. As there are no methods that could estimate or prove those features, any "minimal" solution is only minimal until a better one is found.

Always	Movement_Detected & Dark	Lamp_On Timer_ResetStart
Always	Timer_OVER | Light	Lamp_Off Timer_Stop

Figure 6.6 Security lamp as a combinational system: the table *Always*.

If there are so many solutions possible for a trivial example, we can imagine how many solutions exist for more complex state machines with tens of states. They are surely far away from any minimal number of states. Fortunately, in software the number of states does not play a very important role. We should rather concentrate on easy comprehension of the state machine specification than on its number of states.

The solutions for the Security lamp control problem we have discussed, especially the combinational system, do exist and make sense for a certain execution environment (including that of StateWORKS) but may be unsuitable for other environments.

Models of a Finite State Machine

A finite state machine is also a standard model used in the mathematical foundation of computer science, e.g., in the formal specification of programming languages. Those concepts are, as we describe them in this book, event-driven "Parser" problems. That fact explains at least partly the popular understanding of a state machine in software.

From that perspective a finite state machine (see, e.g., Carroll et al.[1]) a finite state machine is a quintuple $\langle \Sigma, S, s0, \delta, F \rangle$, where:

- Σ is the input alphabet (a finite non-empty set of symbols).
- S is a finite non-empty set of states.
- s0 is an initial state, an element of S.
- δ is the state transition function: $\delta: S \times \Sigma \rightarrow S$.
- F is the set of final states, a (possibly empty) subset of S.

A **parser** state machine is also called a **recognizer** or **acceptor**. Such a state machine has an initial and a final state and its path from the start to the final state is deterministic; i.e., there is only one transition from each state. We would rather say that all other inputs are ignored or by definition they cannot occur. If the state machine, for a given input, accepts several transitions from at least one state, it is called nondeterministic.

Those kinds of considerations are adequate for parsing of words but do not make sense for a control application outside that environment. We note also that that definition misses completely the output (actions). That is understandable as the task of that state machine is to reach a final state, which means that parsing has been successful; there is nothing to do on the way to that state. Theoretically, we could adapt that concept for our purpose — modeling applications implemented by software. It would be required to treat a state machine as an automaton that changes its state, and then by decoding states we decide which actions are to be performed.

That solution would be much more complicated as it is difficult to think separately about state changes and actions: state changes are in most cases the results of some feedback from the controlled application. It is a much simpler and more natural way to construct a state machine if we think at the same time about state changes and about actions to be done.

Staying in the world of symbols, we can define a **transducer** finite state machine as a sextuple $\langle \Sigma, \Gamma, S, s0, \delta, \omega \rangle$, where:

- Σ is the input alphabet (a finite non-empty set of symbols).
- Γ is the output alphabet (a finite non-empty set of symbols).
- S is a finite non-empty set of states.
- s0 is an initial state, an element of S.
- δ is the state transition function: $\delta: S \times \Sigma \rightarrow S$.
- ω is the output function.

If the output function is a function of a state and input alphabet ($\omega: S \times \Sigma \rightarrow \Gamma$), that definition corresponds to the Mealy model. If the output function depends only on a state ($\omega: S \rightarrow \Gamma$), that definition corresponds to the Moore model.

The parser state machine model is useful for compilers. In a design of hardware digital circuits the transducer state machine is applicable.

Application-Based State Machine Models

Several criteria can be used for classification of state machines. The acceptor and transducer (Moore and Mealy) models originated in the educational/scientific world. Taking the application criteria into consideration we would rather speak about parser and controller state machines.

The **parser** state machines match strings (symbols); i.e., they follow a sequence of states with the purpose of detecting a certain string pattern. In practice, they do nothing while changing the states and the outputs (actions) are not used and do not exist for them. We discuss them later in more detail (see *"Parser" problem* in Chapter 7).

A specific type of parser state machines transform changes of input signals into a set of states, which can then be used by simple decoding to determine outputs. Such situations occur relatively seldom and are characterized by very homogeneous control requirements, which happen only for rather simple control systems. We later discuss a state machine TrafficLight (*Example — Traffic light control* in Chapter 9), which represents exactly that kind of model. For the TrafficLight control we use a state machine whose sense is to reflect the position and the movement direction of a train that is in the controlled zone. If we know that information, we can decide about the traffic light: it must be on if at least one TrafficLight state machine signals that the train moves toward the crossing (there is a separate state machine for any train in the controlled zone). The solution is simple and the design of the "sequence" is obvious. Those state machines are limited to Moore models.

Summarizing, parsers are special cases of state machines finding little application in industrial control because there are not too many situations where they may be used.

The **controller** state machines are actual state machines used in practical applications. They consist of states that determine output (actions). With controller state machines, we are able to specify behavior of any complexity, especially using a system of state machines.

A strict classification and distinction between Moore and Mealy models may have value for some theoretical discussions, but from the practical point of view a useful model is a state machine that combines features of both models. That is, we need a state machine where outputs are functions of a state as well as a state and inputs. Such a controller model will be used and discussed in the following chapters.

State Machine Execution Models

State machine definitions are fairly simple in an artificial world of mathematical symbols where the input is a character stream. In the real world the problem becomes more complex. Especially the introduction of outputs, or as we call them actions, makes the problems quite sophisticated.

The state machine model depends on the implementation. We cannot draw something on a piece of paper and claim that the drawing represents the behavior of an application without having in mind the execution model of the state machine. It may be a simple issue if inputs are symbols like characters and we assume that the symbol is an event that, after processing, is treated as consumed and the execution system waits for another event (symbol). If we are in a less homogeneous world where inputs are of varied nature and this simple event-driven model will not work, the situation is not clear at all. Let us try to find an answer to the two basic controversies:

■ A state has several transitions governed by conditions that are complex logical equations. Triggered by an event the execution system finds out that at that moment several transition conditions are due — which transition should be performed?

■ A state has several Input Actions whose conditions are complex logical equations. Triggered by an event the execution system finds out that at that moment several Input Actions conditions are due — which Input Action(s) should be performed? If our model allows several Input Actions to be carried out, does the execution sequence play any role? Can an Input Action be carried out several times (this situation happens if the event has triggered an Input Action but not a state transition)?

The possible answer: "not to use complex logical conditions and limit the model to single symbol dependencies for transition and action" has only an academic value. As we discuss later, such (event driven) state machines (Chapter 7 *Misunderstandings about FSM*) result in a state explosion phenomenon and have no practical value for design of software applications. The answer to the first question could be: yes, allow several transitions (nondeterministic fsm). Again, it may have some theoretical value but it does not seem to have any practical value in industrial control. As there is no clear answer to the above questions, especially to the second, we have several execution models.

> Behavior models must be based on well-established execution models. Otherwise, they are just informal drawings, which may help to explain our intentions to persons who will implement it but the actual design work must be then done in code by programmers.

We take another aspect of a state machine model to show the dependencies of the behavior model on the execution model: let us imagine that for some reason we would like to have a transition to the same state. The reason for such requirement might be to have a chance to repeat the Entry Action. It will work only in some execution systems that assume consuming the condition (event) when it is once used. The usefulness of such an execution environment may be justified, e.g., for a string analysis, but in general it is rather suspicious. If the transition condition is, for example, a value of a digital input, it would be rather strange to pretend that the value can be only once used and then disappears.

In this book we describe various features of the execution model that we assume should be used. This model is not the only one possible, but it is the result of several years of studies, discussions, and experiments. It permits the design of software that is secure and deterministic in its behavior.

Coding as a Universal Solution

The discussed data and control flow separation does not seem to fit well into the concept of OOP, which is the governing paradigm in present-day software development methods. In OOP, objects are the basic elements used for programming. An object should contain both: data and behavior. This view of software development makes it difficult to build software according to Figure 6.1. The misinterpretation of the idea of data and control flow separation stems from the confusion between the application requirements and program implementation. The application requirements define behavior, which can be specified in terms of state machine(s). The implementation of the application requirements is a program that may be written according to OOP rules. We should not mix these two completely different abstraction levels. Of course, if we believe that OOP is a remedy for all problems, we are stuck with it. This book tries to show that an application and its software implementation are two different things — we may use different concepts for their specification.

There are several coding schemes for a state machine. Simple solutions as used in the C language are based on two switches: one for state, another for inputs. Hence, if we deal with a Mealy model (only Input Actions) we could write something like:

```
switch ( state )
{
  case State_0:
  InputAction_0( input );
  break;

  . . .

  case State_N:
    InputAction_N( input );
    break;
  default:
    Error_UnexpectedState();
    break;
}
ChangeState( state, input );
```

where the `InputAction_i()` function contains a switch on input, e.g.:

```
switch ( input )
{
  case Input_0:
    DoAction_0;
    break;

    . . .

  case Input_M:
    DoAction_M;
    break;
```

```
    default:
      Error_UnexpectedInput();
      break;
  }
```

The ChangeState() function changes the state checking the transition conditions for a given state.

For a Moore model we could write something like:

```
  next_state = ChangeState( state, input );
  If ( next_state != state )
    Do_EntryAction( next_state );
```

where the Do_EntryAction() function contains a switch on state, e.g.:

```
  switch ( state )
  {
  case State_0:
    EntryAction_0;
    break;

    . . .
  case State_M:
    EntryAction_K;
    break;
  default:
    Error_UnexpectedState();
    break;
  }
```

In this case, the state is changed first and then the (state dependent) Entry Actions are performed.

Table-Driven Software to Reduce Coding Effort

If state changes can be expressed by a transition matrix, the ChangeState() function can be just a software table as described, e.g., by Wagner.[2] This solution is used for "parsing" state machines, such as the vending machine counter discussed in Chapter 4 (Figure 4.3) whose transitions are described in Table 4.1.

Using C++ constructs the software transition table for the vending machine counter would look like this (where coinAny means the situation which should not happen: a coin other than 5 or 10 cents):

```
enum eState
{
Start, Five, Ten, Fifteen, Twenty, Stop
};

enum eInput
{
coin5, coin10, coinAny
};

eState transition[Stop+1][coinAny+1] =
{//              coin5"        "coin10"      "coinAny"
/*Start*/    {Five,      Ten,          Start},
/*Five*/     {Ten,       Fifteen,      Five},
/*Ten*/      {Fifteen,   Twenty,       Ten},
/*Fifteen*/  {Twenty,    Stop,         Fifteen},
/*Twenty*/   {Stop,      Twenty,       Twenty},
/*Stop*/     {Stop,      Stop,         Stop}
};
```

Each row of the table transition contains "*To*" states for a given state and the input. Thus, the assignment

```
nextState = transition[state][coin];
```

delivers the next state. For instance,

```
transition[Ten][coin10]
```

delivers the value Twenty.

This example shows that in simple cases at least the state transitions can be shown explicitly in a table, which is more comprehensible than a coded alternative.

The complete code of this example (VendingMachine) is in Appendix J *Vending Machine Counter project*. There we show a solution for a State-WORKS execution environment — it will deviate from the simplified view presented in this chapter. The example in this chapter was presented to introduce the presentation means and we have used a simplified specification for that purpose.

Limits of the Coded Solutions

The example above also shows the weak points and limitations of a coded solution. Let us take the state transition table, which if applicable makes the behavior of the state machine comprehensible and easy to modify. The transition table can be used only if the inputs are in a form of ordinal values, typically enumerations or integers. Transition expressions required in practice are as a rule more complex, containing several input conditions linked with Boolean operators. These operators cannot be included in the table,* effectively limiting table usage to a very narrow class of state machines.

There are several other proposals for state machine implementation. Using an OOP approach we may define a state machine as an object that contains both data and the behavior. This theoretically obvious approach is not effective and useful in practice. This approach is also supported by corresponding patterns. To hide the behavior in the object as in a kind of black box may sound interesting but in practice the opposite is expected. We still have the problem of how to implement the behavior in the object. The problem becomes more complicated if we think about a system of many state machines. To have a bunch of objects communicating chaotically among themselves cannot produce a reliably working system, comprehensible and easy to maintain.

Recommended Reading

1. Carroll, J., Long, D., *Theory of Finite Automata with an Introduction to Formal Languages*. Englewood Cliffs, NJ: Prentice-Hall, 1989.
2. Wagner, F., *The Virtual Finite State Machine: Executable Control Flow Specification*. Giessen: Rosa Fischer-Löw Verlag, 1994.

---

* Repeating the "*To*" state is a work-around for the OR operator, but there is no known trick for the AND operator.

Chapter 7

Misunderstandings about FSM

Historical Background

The concept of the finite state machine, although born over 50 years ago, is still not well understood or interpreted in the software domain, despite its wide application in hardware design. Misunderstandings about state machines have produced several stories and half-truths: e.g., the state explosion phenomenon, which discourages the usage of this very useful concept. The concept of the state machine has been several times (unintentionally?) reinvented for software.

A state machine is the oldest-known formal model for sequential behavior, i.e., behavior that cannot be defined by the knowledge of inputs only, but depends on the history of the inputs. The ideas of automaton and states can be found in publications that appeared 50 to 60 years ago. Some examples are cited in the references — Huffman,[1] Mealy,[2] Moore,[3,5] Gill,[4] Kohavi,[6] and Hopcroft et al.[7] In the years since, the name state machine has evolved. In the beginning it was an automaton or a sequential circuit (remember, it was used for hardware design). But the definition stayed, of course, as we defined it already in the previous chapter. The forms of the definition may differ from the graphical one as we use it in this book to purely mathematical as we have shown in the previous chapter.

We may use any form of the finite state machine definition that best suits our purpose. Independently of the form used, the essence of the definition is always the same: it is a model that uses a state to store the past history (changes of inputs).

A state machine is not a heuristic model, which could be interpreted in many ways, at the whim of the user, but rather has a good theoretical basis. This means that it is relatively easy to invent verification methods to prove the correctness of a system built as a system of state machines. Probably, the state machine is the only known model (of the many used in software development) that really gives a designer a chance to verify a control system, and thus, it is the only way to produce reliable control software.

A state machine is intuitively understandable and therefore acceptable to many designers. There are some application domains where the usage of state machines seems natural; telecommunication and industrial control systems are primary examples. Because control tasks are present in all software, state machine models could be used in most software developed nowadays. There are no reasons of principle to avoid the use of state machines. On the other hand, there are no reasonable justifications for believing that to code the behavior of the control system without any formal model could produce better software than a system based on state machines.

The state machine concept is not the only common model. There are other models that have been invented, often inspired by the ideas of states and flowcharts: e.g., the models used in PLC languages. The central idea of all these control models is a state (sometimes renamed). Whether the derived method really implies an improved state machine can be questioned: the reasons for invention of a new method are not always of a technical nature. Anyway, it seems that the state machine remains the preferred model for describing the behavior of control systems.

Software Systems

A state machine was introduced for hardware design and resulted in better organized and understandable hardware projects (in the days when there was no software).

Programming started some time later as a more or less ad hoc activity based on human mental abilities to solve puzzles, rather than on any methods. The problems in software (as programming came to be called) were growing very fast. In addition, the limited number of brilliant programmers required software to be treated as a normal engineering activity with some organization, methods, supervisions, budgets, etc. Among others, the state machine concept started to be used in software design. The integration of a state machine model into software resulted in some new ideas or reinventions.

Some extensions and changes in the state machine terminology have taken place. Especially, the outputs are in software state machines called actions and several types of actions have been defined: entry, exit, input, and transition actions. These changes are welcome and useful extensions, reflecting additional possibilities offered by software implementation. They do not have any relevant influence on the state machine model. They have increased the difficulty of graphical representation of state machines.

Not all ideas have been justified and they may have been caused by missing knowledge. Whatever the reasons were, we should not accept things that do not make sense and often are indeed against common sense. Two typical software issues had a strong influence on state machine misinterpretations and implementations: the "Event-Driven" concept and the "Parser" problem.

Event-Driven Model

Software systems are often event driven. This concept means that the software, in principle, does nothing while waiting for an event. If the event occurs, the software reacts to it and returns to the waiting state. The concept is in opposition to polling systems that continuously check inputs. Conceptually, from the state machine point of view the way of delivering inputs (events or polling) is irrelevant. If the software control system uses a state machine model to describe the system behavior, its model should not be determined by the input acquisition system but rather by the application requirements. Unfortunately, we find in some software implementations of state machines a purely event-driven state machine model, which implies that the state machine has to store not only the history of state *changes* but also the actual present value of inputs.

Parser Problem

This extreme interpretation of a state machine is at least partly explained by the background of some programmers who encountered state machines as a model for Parser behavior. In its basic type a Parser* has one set of

---

* Modern parsers use more sophisticated tools and methods for string analyzing, especially regular expressions allowing whole words to be filtered from strings. We do not discuss here parsers, and the basic idea of character parsing is taken from textbooks.

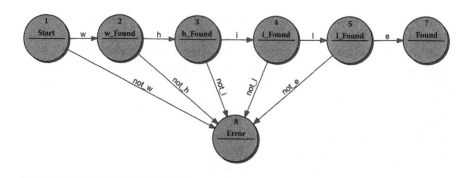

Figure 7.1 The state machine model of a parsing process.

characters as input and detects words in the incoming string. If the Parser has to detect, e.g., a word "while" its specification using a state machine leads to a state machine with few states shown in the state transition diagram (Figure 7.1).

Here, the state really represents the history of (including the present value of) the (single) input. This is correct but such a concept is in practice limited to the parser application. There is no concept of time in this application, and the state machine does not need to know about inputs other than the character stream it is processing. It is no use transferring this idea to other applications as there is no reason to store the present input values in the state because they are explicitly available anyway. The only "reason" would be to make things more complex than they are. We could also mention that a Parser model does not need a concept of actions — it does not do anything; it shows with its state only whether a certain sequence of events has occurred (see discussion in the previous chapter). Although this form of state machine has undeniable usefulness, for example, in lexical analysis for compilers, it has led to essential misconceptions that have retarded progress in the wider software engineering field.

State Explosion

Here we come to the next misunderstanding found in discussions about state machines: the state explosion phenomenon. This phenomenon is directly derived from the above-discussed problem of pure event-driven models of state machines. Of course, if we start to store in the state the present value of inputs, we get the state explosion problem. In this model the number of states increases enormously as each truly required state path must be repeated for all possible input values.

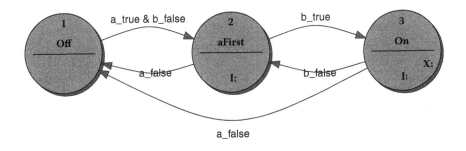

Figure 7.2 Example: the state transition diagram with 3 states.

For instance, let us compare two solutions of a rather trivial example with the inputs: a, b, and c shown in Figure 7.2. The state machine should always set the output Y to `true` if all inputs are `true` and the input a has been `true` before b. The first solution uses a state machine model treating the inputs as they are: values known at any time. The state can have values: *Off*, *aFirst*, and *On*. The transition conditions are indicated directly in the state transition diagram. The combinational part of the state machine that determines the output Y is a function of an input (c) and the state and reads:

Y = On & c

Now consider the equivalent state transition diagram for a purely event-driven version (Figure 7.3). It would require 8 states and very many

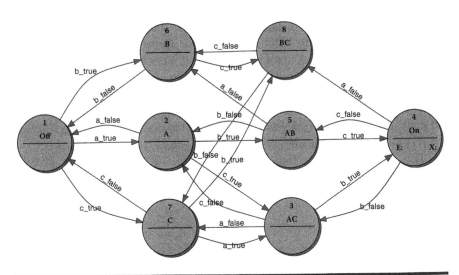

Figure 7.3 Example: the state transition diagram with 8 states.

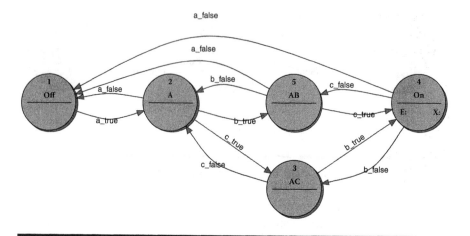

a_false

Figure 7.4   Example: the state transition diagram with 5 states.

transitions. The comprehensiveness of this state machine is of course essentially lower than the first one.

The approach to the event-driven version as presented in Figure 7.3 is of course too dogmatic: it is only to illustrate the idea of storing each event in a state. If we make a more pragmatic design ignoring events b and c in the state *Off*, we still get a state diagram with 5 states shown in Figure 7.4.

This solution could be acceptable as we do not consider 5 (or even 8) states as a state explosion yet (but why use a solution that is obviously more complex than the 3-states automaton?). Imagine the consequences for a more realistic state machine with 6 or more inputs. Any relatively simple problem explodes immediately. And if we want to kill state machine usage completely, make some calculation with 100 or more inputs "proving" that state machines are useless for any practical application.

In *Event-driven software* in Chapter 6 we have seen another example where the event-driven idea requires additional states making the solution more complex without any visible advantages.

The state explosion topic is an example of an artificially created problem that works against usage of state machines.

Signal Lifetime

We round off the topic of event-driven and parser models with a discussion about lifetime (or duration) of signals. As we have mentioned before, when analyzing a string for proving some theorems the problem of time does not exist (the string is present and it is to be parsed). In addition, when parsing is done, the string is treated as "consumed."

In other application domains we cannot consider the input signals in the same manner. Some of them may be consumed after being processed but most of them have a well-defined lifetime. In general, we can distinguish three categories of inputs:

- One-time signal
- Static signal
- Limited duration signal

A **one-time** signal is treated as a true event and is considered "consumed" after being used. Except in the parser we find few examples of such signals.

A **static signal** always has a value. Most inputs are of that sort. For instance, a digital input always has a value (`true`, `false` or `HIGH`, `LOW`). In a more sophisticated implementation, a digital signal may be also `UNKNOWN`, which might be treated as a third (control) value.

A **limited duration signal** has a defined lifetime; i.e.. after some time its value loses its validity. A good example here is a command: its value, e.g., `Cmd_Start`, may be replaced some time later with another command, e.g., `Cmd_Stop`. In such a case it is treated exactly as a static signal. But in contrast to the value of, e.g., a digital signal the value `Cmd_Start` may lose its validity before another value replaces it. For instance, a state machine has started a device, and due to some malfunctions in the system the state machine returns to the initial state. It is obvious that the value `Cmd_Start` must be treated in such cases as invalid; otherwise, the state machine will try to start the device again, which may lead to a loop, which was not the designer's intention. Control software has to provide means for handling the duration of such signals.

State Machine Size

Let us now present the last difficulty: the size of a state machine. Sometimes we read remarks like: a computer is a state machine but the specification of the computer state machine would require millions (maybe billions) of states. Therefore, it would be useless to try to realize a computer using a state machine model. This argument has approximately the same value as a statement like: to write an operating system requires millions of lines of code. Because a programmer cannot see and conquer so many lines it is impossible to program an operating system.

The method of solving complex problems is known: partitioning into smaller units with a good interface among these units. Very often, there is nobody who knows well all parts of a large software system, which was written by several individuals, but the system is in some way comprehensible and conquerable.

This principle applied to state machine specification means that a complex control problem should be partitioned among several state machines, with the entire system of state machines equivalent to a large, single state machine. Note that a pure theoretical calculation shows that, e.g., 10 state machines each having 100 states represents $100 \times 100 \times \ldots \times 100 = 100^{10}$ states. In addition, building of complex control systems does not mean that we start with a complex state machine and then try to partition it. Rather, we specify several state machines for specific tasks and then link them together.

Interface between State Machines

Hence, we arrive at the second topic relating to systems of state machines — the interface among the state machines. Very often we see systems of state machines, which are built as separate units with some communications channel between them which allow any state machine to exchange messages with any other state machine. This is a naive concept based on a programmers' favorite structure: to have a system where each part can send a message to another part. The astonishing fact is that we have known for at least 30 years that this kind of solution leads to deadlocks and similar timing problems. We have even invented nice models like Petri nets to illustrate these difficulties. Anyway, sometimes we forget the basic rules.

To make another comparison — building a system of communicating state machines without any restriction on the communication — is reminiscent of programming with "go to". Sending unconstrained messages to another state machine is like a jump to another program part (Dijkstra[8] showed us 30 years ago the consequences of such "programming style"). Exactly the same problem is generated by unconstrained systems of state machines.

A Flowchart Is Not a State Machine

Flowchart

There are several methods, models, and tools used to describe control systems. Some examples are state machines, Petri nets, Statecharts, flowcharts. Although they describe the same problems, they are not the same thing. Because the state machines are the oldest concept, all methods describing sequential systems are sometimes categorized as equivalent to state machines. This is an obvious misunderstanding, which we discuss further, discussing a flowchart as a counterpart of a state machine.

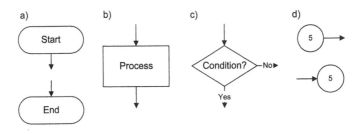

Figure 7.5 Flowchart symbols: (a) Start and End; (b) activity; (c) decision; (d) place marker (to show links between two places of a flowchart).

A **flowchart** is a visual means to show a sequence of some activities. It shows the sequence depending on conditions. Because the wording: sequence, activities, and dependencies remind one of similar terms in state machines, flowcharts are sometimes confused with state machines. In other words, some people do not see the true difference between the two. The confusion starts with the missing understanding of a state, which is not known in a flowchart. To make the matter clear let us first show a simple example.

Flowcharts use a few basic symbols shown in Figure 7.5. Those symbols will suffice for our example as they are enough for most applications. Interestingly, the graphical tool most often used to produce a flowchart offers 26 symbols plus 44 borders and title types (and probably also the possibility to invent a user's own specific symbols, which I have not investigated). It is an interesting example of fascination with graphics. We can understand a graphical representation if we understand the symbol used. Who is able to know the meaning of 26 symbols supported by 44 borders and title types? Nobody, therefore we use only a part of them. If a reader of our drawing knows another subset of the symbols, the reader will not understand our drawing. More does not always mean better.

Example

Let us come to our example. We would like to make a flowchart describing the Security lamp control as discussed earlier in *Event-Driven Software* in Chapter 6, where the details are shown in Figure 6.2 through Figure 6.4. For a reminder we have repeated the state transition diagram in Figure 7.6.

A flowchart that describes the lamp control is shown in Figure 7.7. The flowchart shows step by step the control sequence: the tested condition and the task to be done depending on the test result. We assume that the sequence has to be repeated endlessly: that is, on reaching the

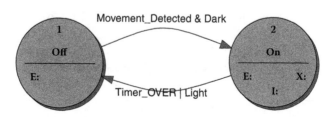

Figure 7.6   Security lamp: the state transition diagram.

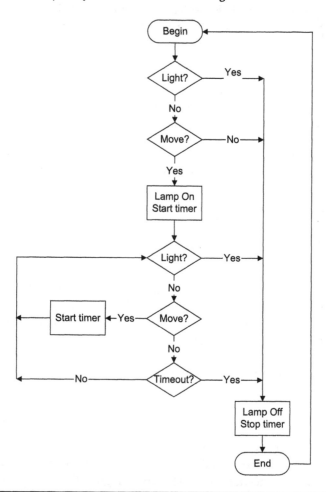

Figure 7.7   The flowchart of the Security lamp control.

end we have to start it again from the Begin (therefore we do without the End symbol). Although a flowchart is intended for sequences that have clear beginnings and ends, nothing can stop us from using it as we

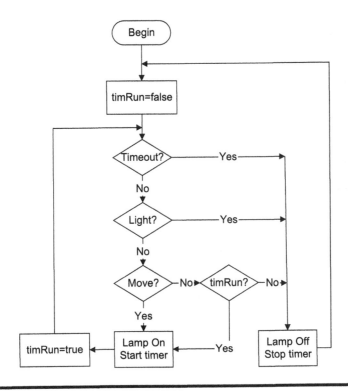

Figure 7.8 Flowchart of the Security lamp control: solution with a flag.

have done in our example. Imagine that we do not like the double existence of "Light?" and "Move?" decision symbols. In other words, we would like to have the flowchart more compact.

If we want to merge the checks, we have to distinguish between the beginning (when the lamp is off) and the situation when the lamp is already on. Then we could decide about time-out checks and actions to be done; in other words we need a concept of a state. We could do it introducing a variable that will represent a state, but then it is not a flowchart anymore but an invention with the same (dubious) value as markers and flags in coded control systems. But we do it and the result is shown in Figure 7.8. We assumed in the second solution that switching on an already lit lamp is not a problem and we use a flag *timRun* to store the information that the lamp is on. If we could get the information from the timer, it would be okay, but if it is not possible we use a flag, and this is a problem. Now imagine that we want to describe not this trivial security lamp problem but some more complex control task where we need several flags (hidden states). Then we end with tens or hundreds of flags and we lose our time (and company money) in endless investigations of why certain flags do or do not have a given value. A flowchart reflects a common practice in coding where flags store information about the past changes of inputs and used resources, such as timers.

In a state machine a state represents the information about the past; if the state machine is, e.g., in a state *On*, we have complete knowledge about the situation: the lamp is on and the timer runs. If we look at a certain point in a flowchart the situation is not clearly defined — we have to know the state of all flags to determine the situation; the flags correspond effectively to a coded state concept.

What Is a Flowchart For?

Flowcharts were initially introduced to describe program flow. An assembler program was really difficult to read, even if well commented. Hence, a flowchart presentation was a help. Even for high-level programming languages it may make sense to describe some program parts in certain situations by means of flowcharts. But also in that case we have to note that they are not a means to describe a program completely. They are a way of getting started, or for documentation, or to explain some concept to students. We cannot expect that a programmer draws a fully detailed flowchart of a major program before coding it, or vice versa, that he prepares a flowchart of a coded program. We are speaking, of course, about software consisting of more than one or two pages of code. We may do it for a procedure but can we imagine thousands of pages of flowchart with cross-references between pages? A flowchart belongs to innumerable means that fake the reality: as if we can really do something serious with it. In fact, many things in software seem to be treated in that way: to have some value for special occasions (presentations), but in the end only code counts.

Obviously a flowchart is a convenient tool. Everybody understands it without any training and it can be drawn without any formal restriction. In other words it is a good tool for informal discussions. It is an illusion to expect more. Of course, there are groups that exploit the concept over any imaginable extent. The procedure is always the same: we intend to use a flowchart for some relatively simple task and after a while we have an uncontrollable monster and nobody knows how to escape from it.

We tend to overuse things that we have learned. We observe that phenomenon when using a programming language. After some time, when we have become highly skilled experts in the programming language we can overuse it. On the one hand we can be proud of our intelligence but on the other hand there can be drawbacks: the constructs of a language are used in a way that was never intended. A good example is the C programming language, which is held responsible for some of the worst software catastrophes. The problem is that the inventors never imagined that C would be used to write such large programs in the way we do. The careless use of pointers that point "unexpectedly" somewhere strange is a primary example.

The flowchart has experienced a similar effect: it is a good tool for certain purposes but it does not mean that it is always ideal for describing a complex control system. And if we use it for that purpose (because we know it), it does not mean that a flowchart is suddenly equivalent to a finite state machine. In fact, we know that flowcharts and state machines are very different things, but we sometimes have difficulty explaining our feeling.

The flowchart concept was a very fruitful idea and can be used in several situations adjusted to specific needs. Although it has lost its common use in code presentations, where it has been replaced by a Program Structure Diagram (called also NS diagram from the authors' names: Nassi–Shneidermann), it obviously influences other methods. For instance, the Specification and Design Language (SDL) is based on a state machine concept but each state is actually described by a flowchart. Another example is the Unified Modeling Language (UML) where the flowchart is used under the name Activity Diagram.

Sometimes diagrams are drawn that are combinations of flowcharts, Petri nets, and state transition diagrams — the IEC 61131-3 Sequential Flow Chart is typical in this respect — and these can be very helpful in documenting intentions, but such unholy combinations can be dangerous to use as they discard any theoretical rigor of their various concepts. They should only be employed for quite simple projects: never for a project that might need to employ tens or more finite state machines in a system.

In summary, a flowchart can be used to describe a sequence of conditional activities. For certain tasks it is a good tool, e.g., to present a product line, an organization, an intention, in general something that we want to do and want to explain to other persons. It is less well qualified to show continuously running sequential activities. The missing concept of state must then be replaced by explicit realization of all imaginable control paths, which is no use except for simple examples in textbooks. By introducing flags to store the past, a flowchart becomes more compact but loses its simplicity and opens the door to dubious inventions, which do more harm than good. Hence, descriptions of control problems with flowcharts are possible but the results are much too complex. And, of course, a flowchart describing a sequence of activities is still a flowchart and not a state machine.

Inventions

After the critical analysis of some incorrect ideas about the state machine concept in software control systems, let us mention some positive developments. The most interesting, true invention has been the introduction of the Statecharts concept. Statecharts have extended the state machine model by elements that make it a completely different modeling tool. The most important element is the view of sub-states through a state. This means that, by definition, a design begins with a single state, which can be expanded into several sub-states, with each of them again expanded into further sub-states. Hence, instead of dealing with several state machines for a complex system we always stay in one state machine expanding the sub-states. To implement this idea Statecharts introduces several additional concepts, some of them quite sophisticated: several state entries (enter-by-history H, conditional C, selection S), several state exits (split, merging by condition, independent), activities and actions, special actions (clear history, internal actions), ORing states, ANDind states, and others. Thus, Statecharts differ from state machines, not only by using rounded rectangles for states instead of circles, but as a truly different way of specifying the behavior of sequential systems.

The Statecharts concept originated in attempts to design a complex avionics system and to define its behavior as a finite state machine. In the course of this work, the state machine transition diagram became remarkably complex, covering a complete room wall, and those involved decided that they had "proved" that the classical transition diagram was impossible to work with! Tragically, they took the wrong direction, having missed the possibility of building systems out of subsystems, or components, so that each was in itself small enough to be manageable. This was perhaps the result of an attempt to extend top-down system design practices, commendable in themselves, to the implementation phase. It seems strange that the lessons of "structured programming" of the previous two decades were forgotten in this instance.

We do not want to discuss Statecharts any further here. We mention it only as an example of an interesting concept that introduces a new model for specifying behavior of a control system. Unfortunately, Statecharts is sometimes considered as a replacement for the state machine concept. These assumptions have either a commercial aspect (UML apostles try to kill anything that does not fit into their world) or just ignorance. There is no direct translation between a system of state machines and a Statecharts representation of a control system. Statecharts is a new model of control systems, which is used in UML for behavior specification and is not too helpful for creating the software.

UML also fell into the "event trap" by assuming a model in which input events that are not immediately useful are discarded, leading to state explosion and partly remedied by the use of "guards," which are inputs governing transitions and actions: piling complexity on complexity and reducing the chance of automatically generating software from totally complete and soundly based models.

We should also like to mention some reinventions of state machines that are true "wheel reinventions" caused probably by their author's ignorance. The authors of this book have seen at least three proposals of this kind, the last where the state machine is called a "transition network." Other elements of a state machine — states and actions — have not been renamed in that reinvention.

Conclusions

We discussed some misunderstandings and misinterpretations that have come into being while applying the concept of a state machine to software problems: the pure "event-driven" state machine model resulting from the "parser" solution, the states explosion phenomenon, and unconstrained systems of state machines. The problems have emerged while some people are trying to "reinvent the wheel" defining anew but incorrectly and unnecessarily the well-known concept of state machines. This chapter is based on a paper written by Wagner and Wolstenholme.[9]

> Continuing, we try to show that the well-known concept of a state machine allows us to formulate a way of designing control systems that do not have state explosion or other disastrous timing and synchronization problems encountered in many systems that use self-invented state machine models. We also show how to eliminate large amounts of complex coding, by use of formal, "Platform-Independent Models" directly in the runtime systems, so as to avoid even automatic code generation.

In our view the ability to create large-scale systems of correctly managed finite state machines working in a well-organized fashion will be of major significance.

Recommended Reading

1. Huffman, D. A., "The synthesis of sequential switching circuits," *Journal of the Franklin Institute,* 257 (1954): 161–90; (April 1954): 275–303. Reprinted in Moore.[5]
2. Mealy, G. H., "A method of synthesizing sequential circuits," *Bell System Technical Journal* 34 (September, 1955): 1054–1079.
3. Moore, E. F., ed., "Gedanken-experiments on sequential circuits," pp. 129–53, in *Automata Studies, Annals of Mathematical Studies,* no. 34, Princeton: Princeton University Press, 1956.
4. Gill, A., *Introduction to the Theory of Finite-state Machines.* New York: McGraw-Hill, 1962.
5. Moore, E. F., ed., *Sequential Machines: Selected Papers.* Reading, MA: Addison-Wesley, 1964.
6. Kohavi, Z., *Switching and Finite Automata Theory.* New York: McGraw-Hill, 1970.
7. Hopcroft, J. E., Ullman, J. D., *Introduction to Automata Theory, Languages and Computation.* Reading, MA: Addison-Wesley, 1979.
8. Dijkstra, E. W., "GOTO considered harmful," letter, 1966.
9. Wagner, F., Wolstenholme, P., "Misunderstandings about state machines," *IEE Comput. Control Eng.* (August–September 2004).

Chapter 8

Designing a State Machine

A State Machine Models Behavior

A state machine is a model of behavior; in general it models a control system that is to supervise an application. This definition means that a state machine is a decision machine that generates signals representing actions: do this or do that. A state machine is stimulated by inputs that represent the accessible knowledge about the controlled application.

Any control problem may have several solutions. Exactly the same control can be achieved by several models of a state machine. Except in trivial cases it is difficult or impossible to prove that one state machine is better than another (what criteria might we use?). Therefore, the dominating factors in discussions about state machine solutions are such things as simplicity, comprehension, maintainability, modularity. A proper design philosophy is the basis for a good solution and allows achieving of these features. Under the title "design philosophy" we understand some rules that a state machine has to fulfill, e.g., interfacing with the outside world (acknowledgment principle), complexity (straightforward or tricky functioning), actions used, completeness of the design. Human factors play, of course, an important role: a given state machine reflects the ability of the designer to translate an informal control specification into a formal logical model in a form of a state machine.

Mealy or Moore Models

Limiting at first the considerations to Entry and Input Actions we are confronted with a decision of which state machine model to use: Mealy or Moore? This is not an academic question because the choice determines our way of thinking while designing the control.

If we prefer to use the Moore model, we tend to orient our thinking strongly on the state. In other words the states dominate our considerations. We think first of all:

> If the state machine goes to the state X it does something (Entry Action) and waits in this state for a reaction to this action.

In this way of designing, a state change is a prerequisite of doing something (of course, the state change is caused by an input change). As a consequence state machines designed according to the Moore model require, in general, more states than others.

If we prefer to use the Mealy model we tend to orient our thinking on the inputs. In other words the inputs dominate our considerations. We think first of all:

> If the input changes the state machine does something (Input Action) and maybe changes a state.

In this way of designing an input change is a prerequisite for doing something, but not necessarily changing a state. The functioning of the state machine is hazier: a link between an input and a state change is not so clear.

At a first glance, the Mealy model seems to be more attractive: the state machine changes a state only if absolutely necessary, handling situations that do not require state changes in the present state. This is not always correct; consider the following example. Let's imagine a state *Waiting* (Figure 8.1), which should be guarded by a time-out (not to be stuck there forever). The state can be reached from several states: *State1*, *State2*, *State3*. Using a Mealy model we have to start a timer in all states that lead to the state Waiting. A more elegant solution is achieved using a Moore model where the timer will be started as an Entry Action in the state *Waiting*. In the Moore model the meaning of the timer is explicitly shown: the timer is started in the state *Waiting* to guard the state.

In Figure 8.1 the Input Action (I) in states *State1*, *State2* and *State3* is StartTimer (see also transition matrix in Table 8.1).

This example shows that the answer to the question "which model is better" is rather subjective, and especially that a term like "elegant" is not

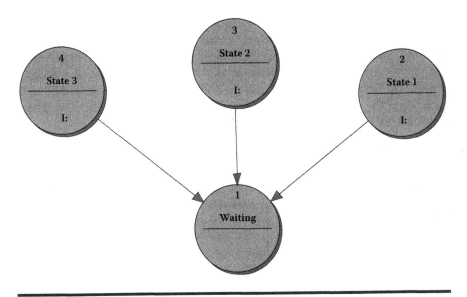

Figure 8.1 Starting a timer in a Mealy model.

Table 8.1 Starting a Timer in a Mealy Model: Transition Matrix

To From	...	Waiting	...
...	...	...	...
State1	...	condition1/StartTimer	...
State2	...	condition2/StartTimer	...
State3	...	condition3/StartTimer	...
...	...	...	...

very precise. Other arguments such as number of states are also not convincing. This point has no great value for a software implementation (a state is just a variable — an integer). It may have some value considering ease of comprehension but experience affirms that a state machine with many, but simple states may be easier to understand than a state machine with fewer, but complex states.

We conclude that the best results are achieved by combining both models: Mealy and Moore. For a hardware implementation this kind of decision is not so simple because it may mean additional expense. For a software implementation the choice of a state machine model has no financial consequences — it influences only the design process.

A combined Mealy/Moore model means that the important actions strongly linked with a state should be specified as Entry Actions. A good example of such an important action is the timer start in the above-discussed state *Waiting*. Auxiliary actions that, as a rule, are not accompanied by a state transition should be realized as Input Actions. This rule underlines the character of the actions: Entry Actions are state dependent, Input Actions are input (and present state) dependent.

Actions (Entry, Input, Exit, Transition)

The classical models of a state machine as defined in Automata Theory know only Entry and Input Actions. In practice, two other actions are imaginable: Exit and Transition. The Exit Action is performed on leaving a state. The Transition Action is a kind of Input Action but is bound to a transition. The question is whether to use them. Also in this case the answer is not quite straightforward. One may think that to have a variety of different actions available makes a design more flexible. This may be true but it does not necessarily mean that a state machine with all imaginable action types will be better. Rather the opposite occurs. The reason is that a design should be understandable, at the initial design stage and also after a while. It should be understandable to be easily modified. It should be understandable so that other persons quickly grasp its functioning. Ease of comprehension can be achieved if the rules of functioning are simple. Therefore, we advocate:

- The use of a Moore model, with Entry Actions, as the leading behavioral path in the state machine.
- Input Actions should be triggered by input conditions that do not cause state changes.
- Exit Actions should be reserved for auxiliary activities that do not result in any reactions of the controlled application. A good example could be here an action that stops a timer when leaving a state.
- Transition Actions are in general superfluous as they do not provide any essential advantages in comparison with other actions. In practice, we do not distinguish between Input and Transition Actions: if an Input Action is followed immediately by a state change, it may be effectively a Transition Action (but not necessarily, as we have shown in a discussion elsewhere in *State transition table* in Chapter 4).

The above rules assume that it is better to operate with fewer action types. In most cases there are no convincing arguments for using all imaginable forms of action. In rare cases the use of more action types may bring some advantages.

Defining States

In a state machine design the choice of the states is probably the most important decision. Theoretically, we should use only those states that are required for a given control (sequential) problem. Automata Theory offers methods that help find an optimal, sometimes minimal set of states required for a given control task. Unfortunately, these formal methods find little application in practice. The first reason is that they are limited, in principle, to the Moore model. The other reason is that the calculation costs become too large for any non-trivial problem. In other words, the Automata Theory methods are good for explaining some rules and ideas but have very limited practical use.

In practice, a designer is responsible for definition of the states. His appreciation of a control problem and his experience dictate to him what states are needed. We should also not forget that — in contrast to hardware implementations — size does not influence the cost of a state machine in software. Therefore, we tend not to care about the number of states. Of course, we should not invent states that are not necessary. We should realize any combinational (state independent) functions without creating artificial states. But we should also not forget that the definition of a state is not so sharp in a software state machine as in classical Automata Theory. A thorough analysis of existing software state machines shows that as a rule state machines have too many states. We observe often that the nature of the sequential problem requires fewer states than are typically used. In most cases we have to accept this situation as a compromise between pure science and reality; in reality we have no incentive to spend time to tune the solution any further.

Acknowledgment Principle (Busy and Done States)

A state machine seldom operates as a single, independent control system. Typically, the state machine communicates with the controlled application and is also a part of a large system of communicating state machines. These dependencies among state machines and controlled application strongly influence the definitions of states. The dependency defined below

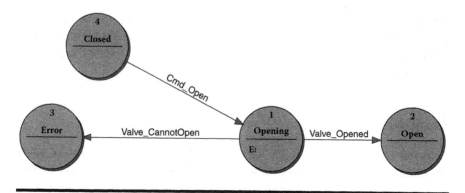

Figure 8.2 Using valve control feedback signal in the Slave state machine.

is known as an **acknowledgment principle**; this is the most important design rule related to the choice of states.

Let us analyze a situation where the state machine Slave performs an action that results in setting an output signal, e.g., opening a valve. If the valve does not have a feedback signal that signals success or failure of the opening action, we cannot do very much. We have to live with a less reliable controller, believing that the action will be successful at any time (of course, the control system will notice it by discovering some undesired effect at a later stage). Unfortunately,

> a controller cannot replace missing information about the situation in the controlled object.

If the valve generates feedback signals the controller should use them as shown in Figure 8.2. Let us assume the state machine Slave is in the state *Closed* and receives (e.g., from the Master state machine) the command Cmd_Open. Entering the state *Opening* the state machine performs the Entry Action Open_Valve. On receiving the positive acknowledgment Valve_Opened the state machine changes its state to *Open*. On receiving the failure signal Valve_CannotOpen the state machine goes to the state *Error*. The valve may generate several different failure answers allowing for more precise evaluation of the situation and corresponding reactions.

This arrangement has a genuine advantage: the state of this state machine signals to the world (e.g., any other state machine) the situation. Its controller (a Master state machine, see Figure 8.3), which has generated the Cmd_Open signal (Entry Action in the state **Busy**), also follows the acknowledgment principle. It sends the command from the state *Busy*

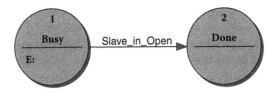

Figure 8.3   Using Slave feedback signal in the Master state machine.

and waits for the reaction of the Slave state machine. If the Slave state machine goes to the state *Open* the master could change to the state ***Done***.

> The acknowledgment principle is a very important rule that should be observed when designing state machines.

Sometimes we forget it or want to save states and try to bypass it by using some tricky solutions. The result is wasted time and an unreliable controller.

The Role of a Timer

A timer plays a very important role in any control application (see the discussion in the previous chapter). The timer is started to count a time period (time-out). When the timer expires it generates an overflow signal. The common function of the timer is to ensure that a state machine does not stay forever in a state.

If a state machine sends a signal to the controlled application we can never be sure that the required action will be performed and acknowledged. For instance, there is always a danger that the communication breaks down for some reason (a cable is broken, a connector falls out, a switch does not work, a data acquisition unit is not in operation). Therefore, it is a "must" to guard states that should wait for acknowledgments from the outside peripherals against this kind of deadlock.

Taking the Slave state machine discussed before (Figure 8.2) we have to guard the state *Opening* with a timer. Therefore, in addition to the Entry Action `Open_Valve` this state has another Entry Action: `Start_Timer`.

Error States and Alarms

Handling of failure situations is a very important point in any control system. In fact, the *sunny-day-scenarios* in a control system are very often not difficult, and to handle them we do not need any formal methods or tools. The true difficulties arise if we want to properly handle all failure situations. Taking into consideration the failure situations increases enormously the difficulty of a control problem. Often the problem becomes so complex that some people believe that any formal methods will fail and they return to "commonsense" solutions (as if the problem might then become simpler). To reduce the complexity we have to distinguish between error states and alarm signals:

> The error state is used to signal to the world that a failure situation has occurred. The alarm signal is information about what failure has occurred and it does not carry any control value.

A simple failure situation has one error source, which means that the alarm signal can be generated in the error state as an Entry Action. In a complex failure situation there are many error sources, but often there is no reason to use a separate state for each failure: in many cases, for the sequential nature of the control process the failure situation is important and not what kind of failure has occurred.

We are going to illustrate this issue continuing the discussion of the state machine Slave (Figure 8.2). Let us assume that there are two possible failures:

- The valve cannot be opened.
- No answer from the valve; detected by the timer.

Instead of introducing two error states we use only one state *Error*, which is reached if in the state *Opening* either the input `Valve_CannotOpen` or `Ti_Over` is true. In addition, the state *Opening* gets two Input Actions:

- `ValveAlarm_CannotOpen`
  triggered by the input `Valve_CannotOpen`
- `ValveAlarm_NoAnswer` triggered by the input `Ti_Over`

We see that the full specification of the state machine becomes more and more incomplete when using the state transition diagram alone (see Figure 8.4). To have all details in one presentation we use the state transition table (we confine ourselves for the moment to the most interesting state: *Opening* shown in Figure 8.5). Note that for completeness we use also the action `Stop_Timer` on exiting the state.

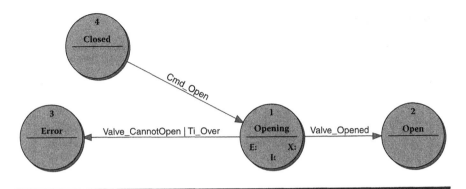

Figure 8.4   Handling of the failure situation by the Slave state machine.

Opening	Entry action	Open_Valve Start_Timer
	eXit action	Stop_Timer
	Valve_CannotOpen	ValveAlarm_CannotOpen
	Ti_Over	ValveAlarm_NoAnswer
Open	Valve_Opened	
Error	Valve_CannotOpen | Ti_Over	

Figure 8.5   The state transition table of the state *Opening*.

Alarms generated by a state machine are information for the user interface and logging files. The error states are control values for other state machine(s) in the system: a Master of the Slave could decide to switch off the valve by sending a corresponding command to the Slave. This example shows the idea behind the tasks of Master and Slave, which we discuss in detail in the next chapter: the Slave handles the details of the failure situation generating (in this case) alarm signals specific for the two failure sources. The Master does not care about the details of the failure receiving only rough information about the failure (Error has occurred) and switches off the Valve control.

Of course, we may handle another control problem where we need to pass to the Master two distinct control values about failure situations by using two error states: *Error_CannotOpen* and *Error_NoAnswer*. This kind of decision depends on the requirements.

Completeness of the Design

A state machine should perform the complete control function, as regards behavior. This is the most important rule for state machine design. If this rule is not kept, the final control system that is partly realized by a state machine and partly elsewhere in the code does not make sense. Unfortunately, coded implementations of control systems tend to ignore this issue, which explains the limited usage of state machines in software. It is probably better to know that the control is hidden in the code instead of living with the illusion that a state machine does the control when it actually covers only a part of the specification. One of the worst sins is to operate with a global variable State, which is then manipulated in any imaginable situation.

Note these rules:

- A state can be changed only in the state machine code.
- A state can be used to perform something only in the state machine code.
- Inputs must not trigger actions bypassing the state machine code (Outputs must not be conditional, being dependent on inputs which bypass the state machine code).

Does this not sound obvious? In practice, we still find deviations, which are justified by the argument that they make life easier. In reality, there is only one reason: it seems so easy to code anything anywhere. If people want to understand a control system ostensibly based on state machines, they will study the specification or implementation of these state machines. They cannot then know that the code contains some other functions that corrupt the state machine.

Let us take the state *Opening* shown in Figure 8.5 and assume that we have done the initial design without taking into consideration the second failure source: no answer from the valve. Receiving the additional requirement we just decide to produce the second alarm somewhere in the input/output handler. After a while we notice that it also requires a state change. So we complete the transition to state *Error* using the input Ti_Over but leave the generation of the alarm ValveAlarm_NoAnswer outside the state machine. Isn't this a typical way of making small changes in the code? This might be considered as a relatively small corruption of the state machine but even this small deviation should be avoided.

Hiding Control Information

> We pray all the time that a specification of the behavior will be comprehensible, i.e., open and clear. This recommendation has limits, which result from the required completeness. In the specification we "see" only the control relevant view of the object, i.e., its control values and actions, while the reasons for their changes may stay hidden.

We will see later that there are both possibilities and reasons to hide control information, e.g., with too complex logical equations, to simplify the appearance of the specification. Hiding some information may make the specification table seem clearer, though it is an illusion — to understand truly the condition we have to see the full definition anyway.

Example — Pedestrian Traffic Lights

The first example is a pedestrian traffic light control. Traffic light controllers are simple systems with a minimal dependency on inputs. In the simplest case a traffic light control is just a state generator (green–yellow–red) without any input. We will take a slightly more complex system with one input: a request button.

The Requirements

We assume that there are two sets of traffic lights. The first set for car traffic contains three lamps: green, yellow, and red. The second set for pedestrians contains two lamps: green and red. Normally, the green lamp for the car traffic and the red lamp for pedestrians are on. The traffic light pillar has a one shot request button. If the button is pushed, the car lights should soon change to yellow. After a while (e.g., 2 seconds) the car lights should change to red and the pedestrian lights to green. After some time (e.g., 15 seconds) the pedestrian lights should change to red. After a time period equal to the yellow phase, the car lights should change to green restoring the normal situation. When the described sequence terminates, there is a disabled phase (e.g., 40 seconds) before the sequence can be repeated. If the button is pushed during the disabled phase, it should be stored and the sequence should be carried out when the disabled phase ceases. The requirements are illustrated with a timing diagram in Figure 8.6. At start-up when the traffic light control is switched

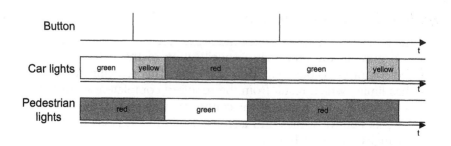

Figure 8.6 **The timing diagram for pedestrian traffic light.**

on, the car traffic light should be yellow for a while, switching then to green. The pedestrian traffic lights should be switched to red from the beginning.

I have formulated these requirements by watching traffic lights installed on a road I have to cross during my daily walks. Other pedestrian traffic lights may differ slightly from these requirements.

The Specification

First of all we have to define (input) control values and (output) actions.

The first input is obvious — it is the control value representing the button; let us call it Di_Request.

A traffic light control uses several timers to measure the required time period. In the discussed case, the state machine needs three timers:

- Ti_Yellow: to measure the car yellow period, and also used for a corresponding red–red overlap period later in the cycle.
- Ti_Green: to measure the pedestrian green period.
- Ti_Disabled: to measure the disabled pedestrian period.

Note that the state machine does not need a timer to measure the car red period as it is defined by the pedestrian green period.

Hence, we have the following control values for the state machine:

- Di_Request
- Ti_Yellow_OVER
- Ti_Green_OVER
- Ti_Disabled_OVER

The state machine has to set traffic lights. Thus, we define the following actions for the car traffic lights:

- C_Red_On
- C_Yellow_On
- C_Green_On

and for the pedestrian's traffic lights:

- P_Red_On
- P_Green_On

By these definitions we have assumed that the electrical part of the traffic lights has some "intelligence" switching on the required lamps and switching off the other ones. In other words, the actions are just commands to a control port, which handles the lamps directly.

The state machine Pedestrian0, which realizes the control, is shown in Figure 8.7. To discuss its functioning we need to see the complete specification as represented in its state transition tables. Three tables are very similar: *Red_Yellow_Init*, *Red_Red*, and *Green_Red*. Therefore, we show only one of them: *Red_Yellow_Init* in Figure 8.8. In all three states some lights are switched on (the state names are chosen so that they indicate the lights *pedestrian-car*) and a timer determines how long the state machine stays in this state.

In the state *Red_Yellow_Init* car lights are switched to yellow and the pedestrian lights are switched to red. The Ti_Yellow timer determines when to leave the state. The two other states build a more interesting

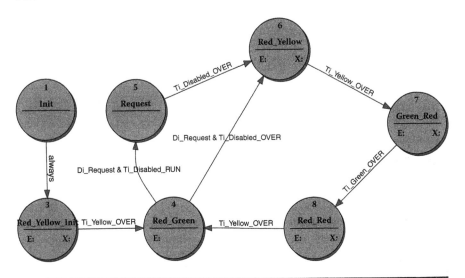

Figure 8.7   Pedestrian0: the state transition diagram.

Red_Yellow_Init	Entry action	C_Yellow_On P_Red_On Ti_Yellow_ResetStart
	eXit action	Ti_Yellow_Stop
Red_Green	Ti_Yellow_OVER	

Figure 8.8 Pedesterian0: the state transition table of the state *Red_Yellow_Init*.

Red_Green	Entry action	C_Green_On Ti_Disabled_ResetStart
	eXit action	
Red_Yellow	Di_Request & Ti_Disabled_OVER	
Request	Di_Request & Ti_Disabled_RUN	

Figure 8.9 Pedesterian0: the state transition table of the state *Red_Green*.

Request	Entry action	
	eXit action	
Red_Yellow	Ti_Disabled_OVER	

Figure 8.10 Pedesterian0: the state transition table of the state *Request*.

group. Their state transition tables are shown in Figure 8.9 (*Red_Green*) and Figure 8.10 (*Request*). Particularly, they handle the Request button. The state *Red_Green* represents the situation when the car traffic rolls having the green light and the pedestrians have to wait. The timer that guards this state does not represent here the time of the car green light but the disabled time during which the control does not react to the Request.

If the Request comes after the timer Ti_Disabled expires, the state machine goes to the state *Red_Yellow* introducing the green phase for the pedestrians. If the Request comes before the timer expires, the state machine goes to the state *Request* where it waits for the timer expiration. If the timer is over, it goes to the state *Red_Yellow* where the car light is switched to yellow. This state is guarded by the timer Ti_Yellow, but in addition, the timer Ti_Disabled must be stopped (it has been started in state *Red_Green* already).

We have so far designed the entire control logic. Now, we have to think about special situations; a start-up is a typical example. A state machine must be set in a defined state after the start-up. This is the Init state in the state transition diagram. We need this state to have the first transition (always) to the state *Red_Yellow_Init* where the initial Entry Action is performed.

The Pedestrian traffic light control is a simple system based mainly on timers. Only the Request input requires some consideration: the presented solution is one of several possible ways of handling this kind of problem. The design is fairly complete; we have tried to take into consideration all information contained in the specification. We will see later that the practical implementation may impose additional requirements (see Appendix K *Pedestrian traffic light project*). For the time being we are satisfied with the result.

The Specification Must Be Understandable

The state machine Pedestrian is simple and really does not present a difficult task to be solved; it requires only an analysis of the sequence of timers to be started and the reaction to the Request button. But let us imagine that we do not bother about signal naming and call the timers in turn T1, T2, and T3 and the states get also rather simple names: *S1* through *S7*. The state diagram and the state transition tables become practically unreadable independently of legend and remarks we put in the project explaining the true meaning of the states and timers. That remark is obvious, is it not? If the naming is such an essential factor for such a simple project, we may imagine its importance in a large one. We should never save time on inventing names and we should correct the names during the project if we find a reason to improve them (of course, the development tools should support the renaming process as there is nothing worse than to change the names by hand).

A behavior specification should solve the problem and must be comprehensible for outside persons who want to see how it works or must continue somebody's project some time later, adapting the control system to new requirements. If we understand our own specification, there is a great chance that other people will also.

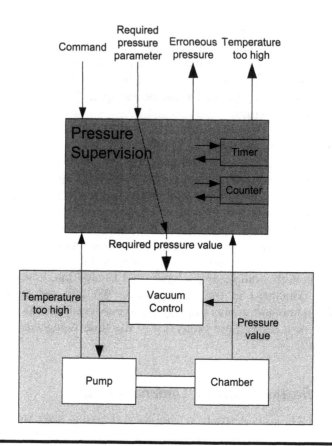

Figure 8.11   A pressure supervision system.

Example — Pressure Supervision

In the second example we design a state machine that supervises pressure in a chamber (see Figure 8.11).

The Requirements

The system to be controlled consists of a *Vacuum Control*, which stabilizes pressure in a *Chamber* by controlling a *Pump*. The task of the *Pressure Supervision* control is to set the *Required pressure value* and to check whether the pressure has reached the required value in a certain time. The pressure value is then continuously checked as to whether it stays in a certain range. Deviations from the allowable range are tolerated if they do not last too long and if the number of such deviations stays below a certain limit; otherwise the *Vacuum Control* should be switched off. The *Vacuum Control* should be also switched off if the temperature of the pump is too high. Any pressure deviation outside the allowable range should be signaled to the operator.

The Specification

We start the design defining the (input) control values and (output) actions. The Pressure supervision state machine has inputs that must contain the following control values:

- Cmd_Start: command
- RequiredPressure_CHANGED: signals that the value has changed
- Pressure_OK, Pressure_TooHigh, Pressure_TooLow: characterize the vacuum in the Chamber
- Temperature_TooHigh: characterizes the Pump temperature
- Timer_OVER
- Counter_OVER

The actions to be performed by the Pressure supervision are:

- SetPressure: sets the value of the required pressure
- SetPressure_Off: switches off the Vacuum control
- Al_PressureError: alarm
- Al_PumpTooHot: alarm
- Timer_ResetStart
- Timer_Stop
- Counter_ResetStart
- Ofun_CalcLimit: calculate limits of the allowable pressure range
- LED_On and LED_Off: switch on and off an indicator of the pressure value

A state machine that realizes the Pressure supervision seems to be not as straightforward as the Pedestrian traffic light control. Let us begin with an initial state, which we call *Idle* (see Figure 8.12). In the state *Idle* the supervision does not work and the state machine waits for a start command. Entering the state *Idle* the state machine switches off the required pressure value.

Let's now introduce a state *Starting* where the state machine goes on receiving the command Cmd_Start. The state *Starting* is a "busy" state where the state machine sets the required pressure and waits for the reaction — if the pressure reaches the required value it will go to the "done" state (which we call *Regulating*). In case of failure: if the pressure

All activities are ceased. Waiting for a Start command.

Idle	Entry action	SetPressure_Off
	eXit action	
Starting	Cmd_Start	

Figure 8.12 Pressure: the state transition table of the state *Idle*.

does not reach the required value within a certain time, it returns to the state *Idle*, also generating an alarm. The state machine also returns to the state *Idle* if the pump temperature gets too high. This consideration leads to a state transition table shown in Figure 8.13.

The Entry Actions contain also a start of a counter and calculation of limits for the allowable pressure range. The range of the allowable pressure must be defined in this moment as this is the base for determining whether the pressure reaches the required value. The counter is started as it should supervise the number of pressure deviations from the allowable range. The last pseudo-Entry Action is clearing the command: we have to do it sometime before returning to the state *Idle* to avoid an immediate restart (see *Signal lifetime* in Chapter 7).

Several activities are initiated. Waiting for Pressure acknowledgements. Due to a Timer missing acknowledgement leads to return to the Idle state.

Starting	Entry action	MyCmd_Clear SetPressure_Set Counter_ResetStart Timer_ResetStart Ofun_CalcLimit
	eXit action	Timer_Stop
	RequiredPress_CHANGED	Timer_ResetStart
	Pump_TooHot	AI_PumpTooHot
	Timer_OVER	AI_PressureError
Regulating	Press_OK	
Idle	Pump_TooHot | Timer_OVER	

Figure 8.13 Pressure: the state transition table of the state *Starting*.

The Pressure value is ok.

Regulating	Entry action	LED_On
	eXit action	LED_Off
	Pump_TooHot	AI_PumpTooHot
Error	Press_TooHigh | Press_TooLow	
Idle	Pump_TooHot	

Figure 8.14   Pressure: the state transition table of the state *Regulating*.

In case of a success (the pressure reaches the required value) the state machine goes to the state *Regulating* (see Figure 8.14). The state *Regulating* is a "done" state where the state machine stays until either the pressure is bad or the pump gets too hot. If the pressure is too high or too low, the state machine goes to the state *Error*. If the pump is too hot, it is of no use for the pressure control anymore and the state machine goes to the state *Idle* to switch off the Vacuum control.

In the state *Error* shown in Figure 8.15 the timer measures the time pressure is outside the range. If this period exceeds its time-out, the state machine goes to the state *Idle*. We assume that we use a counter that counts the number of pressure failures (e.g., the number of entries into the state *Error*). The counter has been started in the state *Starting* and if it overflows the state machine returns to the state *Idle*. Both failures, Timer and Counter overflow, generate alarms before the transition to the state *Idle*.

The Pressure value is outside limits. A return to the state Regulating is possible if the Pressure improves in a certain time. A Counter forces a return to the Idle state if the number of errors exceeds a predefined value (see project settings.) A Timer ensures that a missing acknowledgement will cause a transition to the Idle state.

Error	Entry action	Timer_ResetStart
	eXit action	Timer_Stop
	RequiredPress_CHANGED	Timer_ResetStart
	Counter_OVER | Timer_OVER	AI_PressureError
Regulating	Press_OK	
Idle	Counter_OVER | Timer_OVER	

Figure 8.15   Pressure: the state transition table of the state *Error*.

Specification of state independent actions. If the Required Pressure value changes:
- the Error Counter is reset,
- the Pressure Limits are recalculated,
- the Pressure value is set.

Always	RequiredPress_CHANGED	Counter_ResetStart Ofun_CalcLimit SetPressure_Set

Figure 8.16   Pressure: the combinational part.

If the required pressure parameter changes while staying in the state *Error* the state machine restarts the timer as the Vacuum Control system needs some time to regulate the pressure to the new value. The change of the required pressure parameter may happen at any moment and requires three actions:

- Resetting of the Counter (we assume that after a change of the required pressure value the control functions begin in a sense again from the beginning)
- Calculating of the new limits of the allowable pressure range
- Setting the new required pressure value

We could introduce corresponding Input Actions in all states. Alternatively, we may decide that the control system will have a combinational part, which carries out the actions any time they are triggered by the required pressure parameter change. Hence, the combinational part can be described as presented in the (pseudo-state) table *Always* in Figure 8.16.

The Pressure supervision state machine is more interesting and challenging than the Pedestrian traffic control. Though it is a relatively small state machine considering the number of states, it performs quite sophisticated functions, which were achieved combining the Moore and Mealy models. In the design we took some arbitrary decisions as, e.g., in the definition of control values: Pressure_OK, Pressure_TooHigh, and Pressure_TooLow (instead of Pressure_TooHigh and Pressure_TooLow we could have used Pressure_Bad). The other decision that requires a comment is the use of a counter that counts the number of "errors" (alternatively, we could have used a counter that is incremented explicitly by the state machine on entering the state *Error*).

The Output Function

We have not also discussed how the limits of the required pressure range are used by determining the control values. Until we do speak about an actual implementation we are in fact free from this kind of decision. The main objective of such an implementation-independent specification is the clarity or elegance of the design. If we design a state machine keeping in mind a specific execution model (implementation) we have to take into account limitations and requirements imposed by that implementation. At this stage we are free, not limited by any coding or other restrictions.

In *Data processing result* in Chapter 6, we indicated that results of data processing can be used as control values. In the StateWORKS implementation discussed in Part III data processing units are called output functions and are just C/C++ functions specific to a given application. The function CalcLimits() used in a StateWORKS project for specification of the state machine Pressure is shown in Appendix M *Output function CalcLimits()*.

The State Transition Diagram

For completeness, we also show the state transition diagram of the Pressure supervision state machine in Figure 8.17. In addition to the discussed states the diagram contains the state *Init,* which is the initial state after the start-up. The immediate (always) transition to the state *Idle* guarantees that after the start-up the Vacuum control system receives the signal PressureOff. This might be negligible as the Vacuum control system should be constructed in such a way that it stays off after start-up. If it is the case, the state Idle could be the initial state reducing the number of states to four.

The diagram suggests that the Pressure supervision state machine is a rather simple control problem. The details in the state transition table disclose the true complexity of the machine. This example supports our remark that the number of states is not the only criterion that determines the complexity of the control problem.

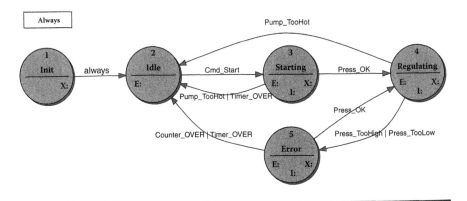

Figure 8.17   Pressure supervision state machine: the state transition diagram.

Conclusions

We formulated in this chapter some rules for state machine design. We bypassed here the typical considerations based on classical definitions found in Automata Theory, which assume that inputs are Boolean values. This limitation does not make sense for a software implementation, which is the actual goal and target of this book. Using the control values for Transition and Input Action conditions and actions for outputs, we are able to specify in an abstract but complete way the behavior of the state machine neglecting details of implementation. This kind of state machine design is not completely implementation independent (see remarks in Pressure supervision state machine) but these dependencies refer to available resources (counter, timer) and, of course, leave open the transformation between the real input/output signals and control values used by the state machine design. This is discussed in a later part of the book.

While designing state machines for the examples we have truly "forgotten" details that are irrelevant to the controller behavior problem. The time periods for the lights, the actual values of the pressure limits, the counter overflow value, the required pressure value: these are all values foreign to the intrinsic design of a controller and as such they are ignored in the discussion. Of course, they will have to be added at some stage, but only as parameters or properties for the "objects," which are the state machines we have designed. For example, the three timer values for the pedestrian crossing could vary, depending on road conditions in various locations, and exactly the same state machine design could be reused in hundreds of situations, suitably configured for each. These values could even be altered at certain times, such as times when many school children were expected.

Chapter 9

Systems of State Machines

Mastering Complexity

A single state machine can control a relatively simple application task, e.g., a motor. Any non-trivial control system should not be designed as a single state machine. Theoretically, we could do it but this approach is not recommended. We have already stressed the requirement that a state machine must stay comprehensible. The limit to the size (number of states) that a designer can effectively develop and understand is not a fixed maximum number of states that a manageable state machine should not exceed. The maximum number of states depends on several factors, e.g., the chosen state machine model or complexity of the application. In practice, the maximum reasonable number of states is between 30 and 100, and some designers are not comfortable when exceeding even 20 states. The larger numbers will be acceptable only if there is an easily appreciated structure to the state transition diagram. Various factors such as the size of the paper or the size of the monitor screen may also play a role here. To understand a state machine we should "see" it at a glance. It gives us at least a feeling for the sequences of transitions.

Normally, applications have several tasks to control. In practice, those tasks are not controlled independently — they form a system in which the parts interact. Thus, to master such an application we should partition the control among several state machines. We need to know how to organize such a system. Should it be a totally free system of loosely coupled

145

state machines or should it have a strict organization? Though there is no ultimate answer to this question we know from experience that a complex system requires an organization. A lesson from software development is that trivial problems are not problems — any solution will do, even spaghetti code — but the difficulties encountered in development of complex systems cannot be solved by *ad hoc* solutions. They require a well-thought-out approach that allows the entire system to be partitioned into several subsystems that communicate smoothly among themselves.

Thus, the way not to get lost is to partition a control task into several smaller, easier to understand control tasks. This approach is characteristic for other domains of design. For instance, we do not write a program as a single, huge 1 million line code but divide it into modules, units, functions, etc., each of which is relatively easy to understand on its own abstraction level.

Partitioning a control task into several state machines has an important side effect; namely, we may operate with a set of standard state machines specialized for certain dedicated functions and used in several applications. This resembles the use of functions or procedures in programming.

Several problems must be solved for a system of state machines:

- The partitioning criteria
- The communication interface among state machines
- The (hierarchical) structure of the control system

The Partitioning Criteria

Speaking about the partitioning criteria, we do not mean: first design a huge state machine and then try to partition it into a set of smaller ones. That may be an interesting academic topic but it does not make any practical sense.

Partitioning is defined by the application that is to be controlled by the system of state machines. Some application domains impose a partitioning strategy; others are not quite so obvious. Partitioning is also a trial-and-error process, which may require some changes or iterations. The first idea is not always the ultimate one.

Consider as an example an industrial process that requires controlling of vacuum, transport, and sputtering, which is typical for the semiconductor industry. There are two partitioning lines for such a process (see Figure 9.1). The vertical lines partition the control task into technological domains, in this case:

- Vacuum control
- Transport control
- Sputtering control

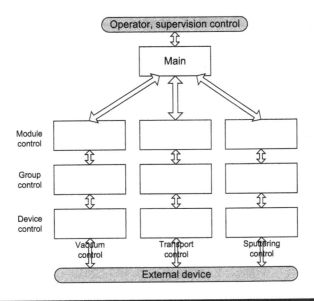

Figure 9.1   Partitioning as defined by the application domain.

The horizontal lines inside each of the control domains partition the system according to abstraction levels:

- Device control
- Group control
- Module control, etc.

Device control covers state machines that control devices on the lowest level; these are state machines that communicate directly with physical devices (pumps, motors, valves, gauges, sources, etc.).

State machines in *Group control* "see" only the device state machines; in a well-designed system they should never communicate directly with the devices, but only through the Device control layer.

Module control is still a higher abstraction, controlling sets of state machines in the group control.

We may need several abstraction levels or not. Sometimes a flat structure may be sufficient. Nearly every system will also require a state machine *Main,* which coordinates the several control domains.

Not all applications are so clear when considering control partitions. Consider, as another example, a system that is to manage a message protocol. The partitioning lines are not well defined in that case. Perhaps, we would consider a state machine that handles the protocol as the core (Main) of the control system. Then, as the needs become evident we will add state machines that each realizes some specific control task, e.g., communication with the modem, communication with the user interface, buffering control, etc.

The sense of the above remarks is to communicate the idea that there are no well-defined criteria that define the partitioning of a control system into a system of state machines. Especially, that the partitioning is strongly governed by communication among state machines and thus by the overall structure of the system.

The organization of a control may require several abstraction levels. The first example above, the example of an industrial control task, illustrates this thesis quite well. The state machine Main, which coordinates the technological branches, is not interested in the states of a single pump or a motor; it can and should work out its decisions on a basis of more abstract control information in a form of "vacuum is ok," "wafer is in position," "sputtering done," etc.

Also in that case, the form of the required abstraction is application dependent and is difficult to express as a set of rules or recipes. Designing a system of state machines cannot be automated; it is a genuine design problem and the result reflects the ability of the designer.

The Communication Interface among State Machines

Building a control system from several state machines means that the state machines must communicate among themselves. The way this communication is organized has a strong influence on the design and implementation.

> This raises a question: What kind of information should state machines exchange? If state machines exchange data, then they would require a message as the information carrier. The data in the message body would contain also control information. But data handling is a different topic, which should not be mixed with the control we are discussing. We do not discuss at this point the software implementation of the state machines but the logical structure of a system of state machines. The concepts presented in this book are based on a total separation between data and control flow, which means that: **There is no case for data exchange among state machines — state machines exchange only control information.**

We have already formulated the desire to have a system where state machines deal with various layers of the control task. Several solutions are imaginable. Speaking of solutions, we are not interested in the transmission

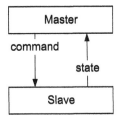

Figure 9.2 Interface between state machines.

or implementation means (messages, events, etc.) but we are concerned with the information exchanged between two state machines. We could discuss interfaces:

■ state–state
■ command–command
■ command–state
■ other signals

We strongly recommend a **command–state** interface between state machines. That recommendation results from our understanding of a relation between two state machines which should be a Master–Slave one: the Master is able to send commands to the Slave and it uses the Slave's states as (inputs) control values (see Figure 9.2). This exchange mechanism has the advantage that it does not require any new concepts to be invented. This arrangement works if

> the state of a state machine represents full information about the control situation in the Slave state machine.

Using the previously introduced terminology we can say that the state of the Slave state machine represents the control abstraction on its level.

Commands issued to a specific state machine are defined by the designer and constitute a set of names (in the software implementation they could be represented by individual integers), as for example

```
{ Cmd_Start, Cmd_Stop, Cmd_Continue }.
```

It is reasonably clear why a state of a Slave state machine should be used as an input for the Master state machine: the state represents the control information about the situation in the Slave. The question of whether to use a state of Master in a Slave must be answered negatively: we should not do it. First of all, it would not be a Master–Slave relation any more. The second, and more important reason is, that by using commands in the way we suggest the Master is given great flexibility in issuing them: it may use the same command in different states or it may generate several commands in the same state. The **Master–Slave** relation means that Master orders the Slave to do something and supervises the Slave by watching its state: if the Slave reaches a certain state, the Master can assume that its command has been carried out. This definition is based on the principle that a state of the (Slave) state machine represents the entire control situation as covered by this particular state machine.

The Handshaking Rule

From the above-formulated requirement (a state of the state machine represents the control situation in the Slave state machine) we can derive the **handshaking** rule, which determines the logic of the communication between two state machines. The handshaking rule requires that a state machine (Slave) uses two states: busy and done to pass the control information properly to another state machine (Master). The explanation of the problem is supported by Figure 9.3, which shows fragments of state transition diagrams of two state machines: Master and Slave. We

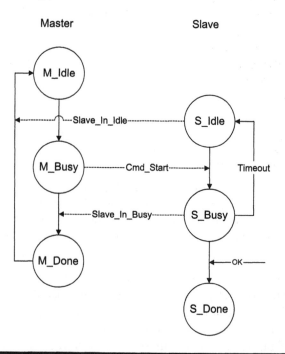

Figure 9.3 The handshaking rule for Master–Slave communication.

analyze three states in both state machines: *Idle, Busy,* and *Done. Idle* is also another done state where the transitions begin.

The Master state machine goes for some reason to the state *M_Busy*; on entering the state it sends a command Cmd_Start to the Slave state machine. That start command causes the Slave to go from the state *S_Idle* to the state *S_Busy*; on entering the state *S_Busy* it sends a command to its controlled device (which might be its Slave). "Seeing" that the Slave is in the state *S_Busy* the Master goes to the state *M_Done*. On receiving an acknowledgment OK from the controlled device the Slave goes to the state *S_Done*. If the OK does not come, the Slave returns (after time-out) to the state *S_Idle*. This allows the Master to return to the state *M_Idle*.

The state *M_Busy* plays here an essential role. Sometimes we imagine that we can do without that state but any trials to eliminate the state busy in Master fail because the condition Slave_In_Idle exists at the beginning if both state machines are in *Idle*. If the Master goes directly from the state *M_Idle* to the state *M_Done,* it will "see" immediately the condition Slave_In_Idle and return to the state *M_Idle*. Of course, we can invent an execution model (see also discussion in the *State machine execution models* section in Chapter 6), which delays sending the information to the Master. We are especially inventive when we code the implementation. The execution model we are propagating (as in StateWORKS, Figure 10.1, the *Vfsm execution model* section in Chapter 10) will cause the infinite loop in the Master state machine.

The (Hierarchical) Structure of the Control System

The interface between state machines suggests that the organization of a system of state machines should have a **hierarchical structure** as shown in Figure 9.4. Each state machine in a system except Main has its Master, which sends commands to it and uses its state as a control value. The state machine Main also has, in fact, its Master, which sends commands to it and monitors its state but this Master is outside the control system we have defined; it may be a remote system, or even a human operator. Similarly, the state machines on the lowest Level 3 may be considered as Masters of controlled (hardware) devices. In a way, each state machine is both a Master and a Slave.

Of course, the Master–Slave interface between state machines does not dictate absolutely a hierarchical structure; it can be used in any system of state machines for exchanging control signals. But in a non-hierarchical structure the role of exchanged control values: commands and states become vague — they contain the information but their interpretation or

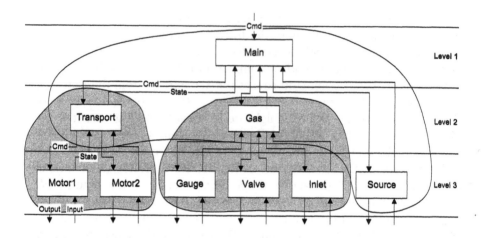

Figure 9.4   A hierarchical system of state machines.

use is not so clear. The entire system becomes hazy and its understanding more difficult.

The recommendation to use a hierarchy of state machines is based on many years of experience and also some considerations of a theoretical nature. When designing state machines, we can make logical errors that could result in infinite loops or deadlocks. The more state machines are in the system, the higher is the probability of such errors and finding of those errors is more difficult. Imagine a system of 100 state machines where each state machine may send a command to any other state machine and uses states of any state machine for definition of its behavior. It is very difficult to understand how such a system really behaves and the "structure" of such a system corresponds to non-structured software.

In contrast to a non-structured system a hierarchical system is easier to design and understand as in its design or debugging we usually solve several local problems: both for a Master and for its Slaves.

Figure 9.4 shows a hierarchical system with three levels: the upper level (Level 1) containing a single Master state machine (Main), the second level (Level 2) containing two state machines (Transport and Gas), and the lowest level (Level 3) with six state machines (Motor1, ... Source).

Design Procedure

Locality of the Control Problems

A design of any state machine always requires an analysis of a limited part of the system:

- A design of the state machine Motor1 covers the outside signals and commands from the state machine Transport.
- While designing the state machine Transport we are interested only in the states of state machines Motor1 and Motor2 and commands from the state machine Main. The state machine Transport has no idea about the details of the Motor1 and Motor2 control (inputs, outputs, delays, time-outs, parameters, etc.); it "sees" only the (abstract) situation in Motor control, represented by the Motor1 and Motor2 state.
- While designing the state machine Main we take into account only states of its Slaves (Transport, Gas, and Source).

The three Master–Slave systems are presented in Figure 9.4 with different shadings covering the areas of the three Masters: Main, Transport, and Gas.

In a correctly designed hierarchical system any state machine "sees" only the states of its Slaves that represent an abstract — but all the same complete — set of the control signals relevant for the state machine. The only state machines that "communicate" with external signals are the state machine Main and the state machines on the lowest hierarchy level.

A hierarchical system allows us to think locally when designing state machines, seeing only the few involved state machines (Master and its Slaves) in an abstract way.

Up-Down or Bottom-Up Design

The other problem is how we should proceed in the design of a state machine system. There is also no strict rule in this case, but it is obvious that we ought to start where the information is the most complete. As a rule we have relatively good information about the behavior of the state machines on the lowest level: how to control peripheral devices. We probably also know about commands for the state machine Main, but unfortunately we do not know much about its Slaves. Therefore, in most cases the only reasonable way is to start with the lowest level of state machines. Having this layer ready, we get some idea how to organize the Master layer above it. So, we continue until we reach the state machine Main. These considerations do not preclude the preparation of an initial plan by a top-down design process, but the finally implemented structure will often depart from the initial plan.

The next question is the design of the hierarchy: how many layers and which state machines are needed. There are some fixed points, e.g., we need a state machine Main at the top and we know very soon which state machines are required at the lowest level — the lowest level is well defined by the controlled devices. The definition of the rest is a process that is strongly bound to progress in designing the lowest level of state machines. Building a hierarchy is an evolutionary process with several trials and it does not have an ultimate solution. The solution reflects the designer's ability and preferences: some people like to use a few, rather complex state machines; other people prefer many simpler state machines in an elaborate hierarchy. There is no definite answer to the question of which approach is a better one, as long as the designer can be quite certain that he or she fully comprehends the way each state machine will function, in both normal and abnormal situations.

Deadlocks

Discussing the design of a state machine we underlined the role of timers, which are very important elements guarding state machines against dead-lock. Any situation that requires an acknowledgment of actions by an external signal must be protected by a timer.

In a well-designed system of state machines only the state machines that communicate with the outside world should use such timers. If we encounter a situation where we need a timer as a safeguard element in a state machine that has no direct link to the outside world, it is a strong indication that there is a failure in the design of a state machine that communicates with the I/O system. In a hierarchical system only the lowest level of state machines uses timers for this purpose.

This rule does not exclude the use of timers for other purposes in any state machine. For example, timers might be used to define time steps in a control sequence (see the *Example — Pedestrian traffic lights* section in Chapter 8).

Loops

A state machine may go into infinite loop sequencing continuously through some state. This may happen when the transition conditions are fulfilled at the same time in all states of a certain control path. It is a rather unpleasant situation, difficult to debug.

The situation can be much more difficult in a system of state machines. A loop that closes through several state machines is essentially more difficult to debug and remove than a loop in a single state machine.

Therefore, a state machine development environment has to provide a (logical) step mode where the step is, e.g., one state transition. In addition, the execution environment of a control system has to provide some mechanism to detect infinite loops and guard the runtime system against the side effects of such loops (processor overload).

Sins

We do not like to impose any strict barriers considering the structure of the system. Any pigheaded, inflexible structure will be rejected in practice as unworkable. Hence, we may design any system, e.g., having in principle a hierarchical system but with additional links between some state machines. How much deviation from a hierarchical system could be tolerated? The diagram in Figure 9.5 shows two examples of sins committed by a system designer.

The undesirable links Cmd1 and Cmd2 are shown in bold. In case Cmd1 a true Master (Transport) of the state machine Motor2 is bypassed, as the state machine Motor2 receives commands directly from the state machine Main (in addition to the commands from the state machine Transport). It is very difficult to understand the behavior of such a system. The probability of malfunctioning is very high as the true Master of the state machine Motor2 (Transport) does not "know" about the additional command. The additional commands are in fact a kind of unexpected input that must be "corrected" by the Master (Transport). On the other hand, if we consider the influence of these additional commands in design of a Master state machine we find no reason for them: those commands could be passed directly to the true Master (Transport). Similarly, it is difficult to design a state machine like Source, which gets commands from Main and also from Gas (Cmd2). What kind of design philosophy to follow?

A similar corruption would be to use a state of the Motor2 in conditions in Main or a state of Source in Gas. All such bypassing is a set of tricks that may give us an illusion of a simplification of a Transition or Input Action condition. We pay for such illusions inevitably with increased errors,

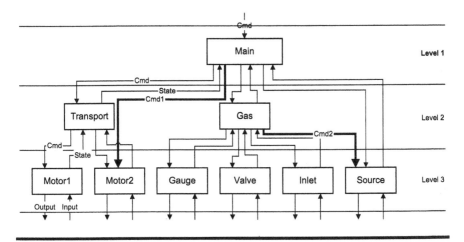

Figure 9.5   A corrupted hierarchical structure.

which are caused by difficulties in understanding system behavior: the system is becoming more and more mysterious with each additional trick.

There are many possibilities to corrupt the hierarchical system, similar to any other software corruption practice. We should avoid them as in general there is no real need for them. It is better to put in some additional effort and make a correct design, avoiding any non-hierarchical deviations in the system, instead of succumbing to the illusion that doing it quickly but wrongly will save time.

Design Rules

Our recommendations for building a system of state machines are as follows:

> ■ Use Master–Slave (command/state) interfaces for communication between state machines.
> ■ Organize the system as a hierarchy of state machines based on a Master–Slave principle.
> ■ Use bottom-up design.
> ■ Avoid corruption of the hierarchy by "wild" links among state machines.
> ■ Use timers guarding against deadlocks only in state machines that communicate with the external signals.

Using these rules we are able to build reliable and maintainable control systems with a high degree of reusability. Well-designed state machines can be stored in a library and used in future projects. The reuse of state machines applies especially for state machines on the lowest level, which control peripheral devices.

The rules are just guidelines that should be followed in most situations. In specific, well-founded cases, we may deviate from them. We show here two designs of a system of state machines. The first is a typical hierarchical system; the second is a non-hierarchical system.

Example — Pumps Supervision System

Task Definition

Consider a control system that is to supervise two Pressure controllers as discussed in the *Example — Pressure supervision* section in Chapter 8. We will call the system Pumps (see Figure 9.6). In addition, it should switch on a Device if the vacuum is okay. If the vacuum fails, the Device

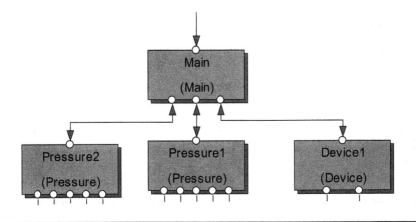

Figure 9.6   The Pumps system.

should be switched off. The Device is also switched off in case it is not acknowledged appropriately. The details of the state machine Device are shown in Appendix L *Pumps supervision project.* The entire system receives commands: Start and Stop from the operator. In line with the preceding discussion, these commands will be issued to the state machine Main. The command Start starts the process (switching on the supervision of pressure and automatic switching on Device if the vacuum is okay). The command Stop switches off Device.

The Pumps supervision cannot be switched off; once started it runs until the vacuum fails (normally, switching off pumps will result in a bad vacuum, which switches off the pumps supervision).

There is only one state machine left in the system to be designed: Main. If the Slave state machines are properly designed (i.e., their states contain full control information about the situation), we need to "see" only the state transition diagrams of Slaves while designing the state machine Main. For convenience we repeat the diagrams here (Figure 9.7 and Figure 9.8).

The First Approach

Up to now, we could define two control values for the state machine Main; corresponding to the two commands we have defined: Cmd_Start and Cmd_Stop. We will introduce other control values on demand: we know that those will be states of Slaves.

Let us make the first trial (see Figure 9.9). We start the design with a state *Idle* where the state machine Main waits for the command Cmd_Start. Next we define the state *StartingPressure*. Receiving the

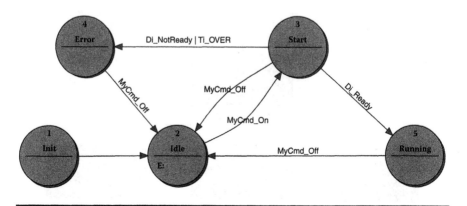

Figure 9.7 Device: the state transition diagram (repeated from Appendix I: *Pumps Supervision Project*).

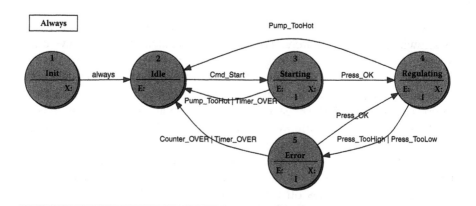

Figure 9.8 Pressure: the state transition diagram (repeated from Chapter 8, *Example — Pressure Supervision* section).

Cmd_Start the state machine goes to the state *StartingPressure* and sends from there the commands PressureCmd_Start to both state machines Pressure. In the state *StartingPressure* the state machine just waits until both state machines Pressure are in the state *Regulating*, which forces Main to go to the next state we need, namely, *StartingDevice*. In the *StartingDevice* state Main sends the command Cmd_On to the state machine Device and waits until Device goes to the state *Running*, which forces Main to go to the state *On*.

The state *On* is the "done" state of Main: the Cmd_Start, which has initialized this series of transitions and actions, has been completely carried out. Main stays in this state until it receives the Cmd_Stop, which forces it to go to the state *Stop*, which is the last state we need. Detecting the acknowledgment from Device (when it is in the state Idle) Main returns

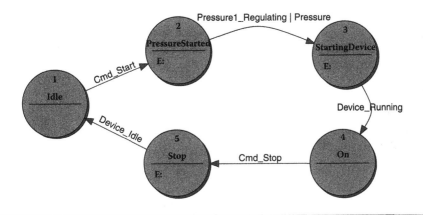

Figure 9.9 Main: the first (erroneous) design.

to the *Idle* state. In such a way we have specified the "normal" control path in the state machine Main when everything goes as we wish. We may say that we have realized the *sunny-day-scenario*. It seems to be not too complicated. Unfortunately, it is not only incomplete but also contains one essential error — we will see it by going into fine details of the control task.

In the design process we have defined so far for the state machine Main, the following control values:

- Cmd_Start and Cmd_Stop, which represent commands from the operator
- Device_Idle and Device_Running, which represent states of the state machine Device
- Pressure1_Regulating and Pressure2_Regulating, which represent states of the state machines Pressure (we call the state machines Pressure1 and Pressure2)

and the following actions:

- DeviceCmd_On and DeviceCmd_Off, which represent commands to the state machine Device
- Pressure1Cmd_Start and Pressure2Cmd_Start, which represent commands to the state machines Pressure

The Second Trial

Unfortunately, things may go wrong (remember Murphy's law), which makes the control task more complicated. In the next round we have to

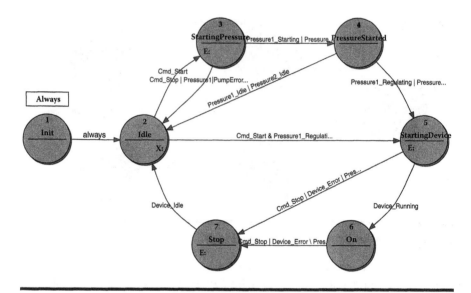

Figure 9.10   Main: the state transition diagram.

analyze the situation in each state trying to foresee all difficulties that may occur and introducing the appropriate reactions to them. After several trials we designed a state machine Main for which the state transition diagram is shown in Figure 9.10. In the following discussion we try to explain how we have come to this solution.

State Idle (see Figure 9.11): the decision to go to the state *Starting-Pressure* when receiving the Cmd_Start might be unnecessary if the

The state machine waits for command Start.
Receiving the command it by-passes the pressure initialization if the pressure is already ok going directly to the state StartingDevice, otherwise it goes to the state StartingPressure. Note the transition priorities.
As the command is not used later it is cleared on leaving the state.

Idle	Entry action	
	eXit action	Cmd_Clear
StartingDevice	Cmd_Start & Pressure1_Regulating & Pressure2_Regulating	
StartingPressure	Cmd_Start	

Figure 9.11   Main: the state transition table of the state *Idle*.

state machines Pressure are already in the state *Regulating* (remember that once successfully started the state machines Pressure stay in the state *Regulating* until the pressure falls out). In such a case, it would be correct to go immediately to the state *StartingDevice*.

The function of the original state *PressureStarted* is now realized by two states: *StartingPressure* and *PressureStarted*. We wanted the state machine Main to wait in the state *PressureStarted* until both state machine Pressure1 and Pressure2 reach the state *Regulating*. If the state machine Pressure cannot reach the state *Regulating*, it will return to the state *Idle*, which should force Main also to return to the state *Idle*. If we add this transition in this state, the state machine will not work properly: at the instant Main enters the state *PressureStarted* the state machines Pressure are still in *Idle*. We assume that the state machines work in sequence, i.e., if the state machine Main terminates its activity, the Slave machines execute.

Taking into account the priority rules for transitions we put the transition to the state *StartingDevice* in front of the transition to the state *StartingPressure*. As the Main command has to be "consumed" before returning to the state *Idle*, we add the appropriate Exit Action here.

State StartingPressure (see Figure 9.12): After sending the command Cmd_Start to state machines Pressure, Main waits, in principle, until

An intermediate state where the state machine waits for Pressure slaves acknowledgments to the commands Start sent to them on entering the state.
Receiving positive acknowledgements: Starting or Regulating it goes to the state.
PressureStarted, otherwise it returns to the state Idle. Receiving Error states from slaves it acknowledges that sending the command Break to the slaves.
Note missing timer - the slaves are responsible for that, the Main master control is based exclusively on slaves' states.

StartingPressure	Entry action	Pressure1Cmd_Start Pressure2Cmd_Start
	eXit action	
PressureStarted	(Pressure1_Starting | Pressure1_Regulating) & (Pressure2_Starting | Pressure2_Regulating)	
Idle	Cmd_Stop | Pressure1_PumpError | Pressure2_PumpError	

Figure 9.12   Main: the state transition table of the state *StartingPressure*.

It is a busy state where Main waits for Pressure slaves performing the required operations: achieve the proper pressure.
If the slaves reach the goal the state machine goes to the state StartingDevice, otherwise it returns to the state Idle sending the command Break to slaves.

PressureStarted	Entry action	
	eXit action	
StartingDevice	Pressure1_Regulating& Pressure2_Regulating	
Idle	Pressure1_Idle | Pressure2_Idle	

Figure 9.13   Main: the state transition table of the state *PressureStarted*.

Pressure is in the state *Starting*. But it is imaginable that one of the state machines Pressure is already in the state *Regulating*. Hence, we have to expand the transition condition to `Pressure_Start OR Pressure_Regulating`. For completeness, we have to foresee a break with the `Cmd_Stop` which forces Main to the state *Idle*.

State PressureStarted (see Figure 9.13): Main waits for the state machines Pressure until they go to the state *Regulating*. If the state machines Pressure have problems, their timers will force them to the state *Idle*. If either of the state machines Pressure goes to the state *Idle*, then Main breaks the process returning to its state *Idle*.

State StartingDevice (see Figure 9.14): in comparison with the initial design the state has been only complemented with the transition to the state *Stop*. We see that there are several reasons to go there: `Cmd_Stop` and problems in Slaves. We note also that in that case there is no acknowledgment problem with Device: in case of difficulties the state machine Device goes to the state *Error* and not to the usual state *Idle*. Hence, we do not need a state *DeviceStarted* comparable to the *PressureStarted* state. This situation shows the coupling between Master and Slave designs: different behavior in Pressure in case of an error in starting could eliminate the state *PressureStarted* in Main.

State On (see Figure 9.15): in comparison with the initial design the state has been only complemented with the transition to the state *Stop*. There are two reasons to go there: `Cmd_Stop` and problems in state machines Pressure.

State Stop (see Figure 9.16): there are no changes here in comparison to the initial design — if Device goes to the state *Idle*, then Main returns to the state *Idle* too. This state demonstrates the difference between internal

A busy state: the state machine waits whether the Device slave is running. Receiving that information it goes to the state On. The control process may be interrupted by slaves signalling problems or the command Stop: in such a case the state machine goes to the state Stop.

StartingDevice	Entry action	DeviceCmd_On
	eXit action	
On	Device_Running	
Stop	Cmd_Stop | Device_Error | Pressure1_Idle | Pressure2_Idle	

Figure 9.14   Main: the state transition table of the state *StartingDevice*.

This is a final state where everything works as required: the pressure in the two chambers controlled by the Pressure slaves is ok and the Device slave is running.
Disturbances in the slaves or the command Stop cause the state machine to go to the state Stop.

On	Entry action	
	eXit action	
Stop	Cmd_Stop | Device_Error | Pressure1_Idle | Pressure2_Idle	

Figure 9.15   Main: the state transition table of the state *On*.

The state is only for sending a command Off to the slave Device.
Receiving the acknowledgement from the Device the state machine goes to the state Idle.

Stop	Entry action	DeviceCmd_Off
	eXit action	
Idle	Device_Idle	

Figure 9.16   Main: the state transition table of the state *Stop*.

and external conditions. Main sends a command `Cmd_Off` to Device. The state transition table of Device shows that it is the only signal needed for going to the state *Idle*. So, we assume that it will happen and do not use a timer to guard Main against a deadlock here. (If it does not happen, it means that the execution software is corrupted and the timer cannot help here as it will not work either).

We note that there has been no demand for using timers as safeguards in any state of Main. The Slaves — Pressure1, Pressure2, and Device — guarantee that the state for which Main is waiting will be always reached in an intended time. In other words, as we have designed the Slaves behavior (Pressure and Device) without any danger of deadlock, Main is safe and does not need timers.

The Ultimate Solution

Though the system that we have designed so far seems to work, we still have a problem. We have left it to the end to show now that there is a difference between a design of a single state machine and a design of a state machine that has to be used in a system. We know that state machines "communicate" using the command–state interface: there is no way to pass control information in another way (more precisely, we can of course do it but we should not do it).

Imagine a situation where the pump is too hot and switched off and the Main sends the command `Cmd_Start`. The state machine Pressure will go to the state *Starting* and return immediately to *Idle*; effectively, from the Main point of view it will stay in the state *Idle*.

In that moment we should make a break and mention the assumption we have in mind, an assumption bound with the execution model of a state machine (see the *State machine execution models* section in Chapter 6). It would be possible (especially if the specification will be coded) that we manage to realize a different execution sequence: after sending the command from Main to Pressure to stop the execution of the Main, start the execution of Pressure, after a state change in Pressure — stop its execution, pass the information about a state change in Pressure to Main, and resume the execution of Main. Our aim is to use a standard execution environment and not to code control sequences depending on a specific situation. The Vfsm execution environment we use (Figure 10.1, *The Vfsm execution model* section in Chapter 10) realizes control sequences according to the following rule: state machines are executed in sequences; once started, the execution of a state machine is continued until there are no state changes due.

When designing the state machine Pressure we had not foreseen passing this information to its Master (remember, we have designed the state machine not thinking at this moment about its usage in a system).

In such a situation Main remains in the state *StartingPressure* until the operator sends the command Cmd_Stop. This may be acceptable but it is not perfect: we wanted to have an automatic control process meaning that after sending the Cmd_Start the system either reaches the state *On* or returns to *Idle* informing us about the reason of a failure.

Trying to improve the design we may consider several solutions:

- Add a timer to Main in the state *StartingPressure*: a bad design, we have advised against the use of time-outs in Masters.
- Use the Pump_TooHot signal in Main to prevent starting the process: a very bad design (see the *Sins* section above).
- Replace in the state *StartingPressure* the transition condition to the state *Idle*: Pump_TooHot OR Timer_OVER by Timer_OVER: this will work but unnecessarily delays the return to *Idle* (and in the meantime the pumps will burn out).
- Add another very short time-out in the state *Starting* (a variant of the previous solution): it will cost us another timer.

Perhaps, there are still more solutions but probably the only correct one that is in line with our recommendations (pass control information using the state) is to add a state *PumpError* in Pressure (see Figure 9.17). If the pump is too hot, Pressure leaves the state *Starting* or *Regulating*, going to the state *PumpError* where the corresponding alarm is generated and waits there for a Cmd_Break (we need another command), which forces Pressure to the state *Idle*.

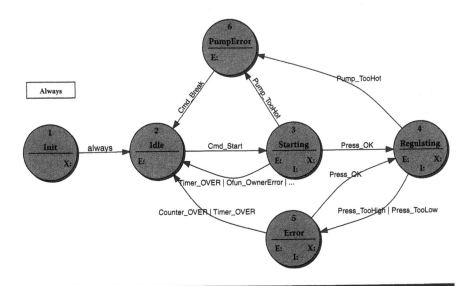

Figure 9.17   Pressure: the ultimate state transition diagram.

As we need the Cmd_Break also in other states (*PressureStarted* and *On*) we may move it to the table *Always* (as in the complete project shown in Appendix L *Pump supervision project*). The changed state Starting is shown in Figure 9.18 and the new state *PumpError* in Figure 9.19. Similarly, in the state *Regulating* the direct transition to the state *Idle* is replaced by a transition to the same *PumpError*. The state machine Main must be also slightly changed: we add two Input Actions in the state *StartingPressure* (see Figure 9.20).

Several activites are initiated. Waiting for Pressure acknowledgments. Due to a Timer missing acknowledgement leads to return to the Idle state. Too hot pump leads to the PumpError state. Both erroneous situations generate corresponding alarms. Error in accessing the output function returns the state machine also to the state Idle: it does not make sense to supervise the pressure without having proper pressure limits (corresponding alarms are generated in Always table).

Starting	Entry action	MyCmd_Clear SetPressure_Set Counter_ResetStart Timer_ResetStart Ofun_CalcLimit
	eXit action	Timer_Stop
	RequiredPress_CHANGED	Timer_ResetStart
	Timer_OVER	Al_PressureError
PumpError	Pump_TooHot	
Idle	Timer_OVER | Ofun_OwnerError | Ofun_ParameterError	
Regulating	Press_OK	

Figure 9.18 **Pressure: the state transition table of the corrected state *Starting*.**

PumpError	Entry action	Al_PumpTooHot
	eXit action	
Idle	Cmd_Break	

Figure 9.19 **Pressure: the state transition table of the state *PumpError*.**

An intermediate state where the state machine waits for Pressure slaves acknowledgements to
the commands Start sent to them on entering the state.
Receiving positive acknowledgements: Starting or Regulating it goes to the state
PressureStarted, otherwise it returns to the state Idle. Receiving Error states from slaves it
acknowledges that sending the command Break to the slaves.
Note missing timer - the slaves are responsible for that, the Main master control is based
exclusively on slaves' states.

StartingPressure	Entry action	Pressure1Cmd_Start Pressure2Cmd_Start
	eXit action	
	Pressure1_PumpError	Pressure1Cmd_Break
	Pressure2_PumpError	Pressure2Cmd_Break
PressureStarted	(Pressure1_Starting | Pressure1_Regulating) & (Pressure2_Starting | Pressure2_Regulating)	
Idle	Cmd_Stop | Pressure1_PumpError | Pressure2_PumpError	

Figure 9.20   Main: the state transition table of the corrected state *StartingPressure*.

To make the state *Starting* of Pressure perfect, we also expanded the
transition condition to the state *Idle*. The transition is due not only when
the timer expires but also when the Output function signals errors. The
Output function calls the `CalcLimit()` function to calculate the switch
point limits: if the calculation fails it is of no use to continue and the state
machine returns to the state *Idle*. Corresponding alarms are generated in
the table *Always*.

We manage in Main without an additional state: any handshaking in this
solution does not make sense as the Pressure transition will be done for sure.

The last changes improved the design. They showed us also that a
state machine that operated satisfactorily as a single controller was not
sufficiently well designed for an integration in a system of state machines.
We stress with the last corrections again the basic rule of design of a
system of state machines:

> The control information from Slaves to Master should be passed
> by states only.

The implementation of the Pumps projects in StateWORKS is shown in
Appendix L *Pumps supervision project.*

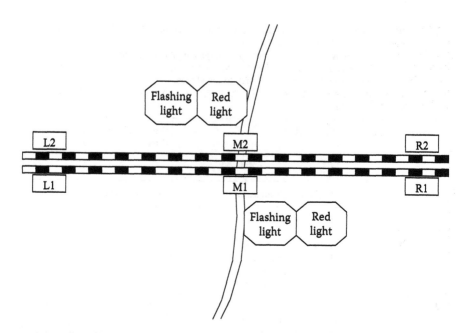

Figure 9.21 Traffic lights at a level-crossing of a railway and a road.

Example — Traffic Light Control

Task Definition

We now discuss an example that is a completely different design in comparison with the previous one. The main difference lies in its structure: it looks like a hierarchical system but in reality there is no hierarchy there.

The design is a traffic light control at a level-crossing of a railway and a road. The control rules are well known but as there may be some variants let us set out the details.

The railway consists of two tracks. Each is monitored by three sensors L (left), M (middle), and R (right) — see Figure 9.21. We assume that a train may come from either direction on each track, but only one train (per track) can enter the sensor zone (we call the space between sensor L and R the *sensor zone*). Further, we assume that a train never changes its movement direction. We assume also that trains may be short or very long; i.e., a train in the sensor zone may cover no sensor, one sensor, two sensors, or all three sensors at the same time.

The control system should cover all imaginable situations, switching the traffic lights according to the following rules:

- Both lights are switched off if the control system is not working.
- The yellow lights are flashing if there is no train between sensors L and R or a train just passed the sensor M and is still between sensors M and L (R) moving toward R (L), i.e., leaving the sensor zone.
- The red lights are on if there is a train between sensor L (R) and M moving toward M, i.e., approaching the road.
- Both red and yellow lights are on if the situation is not definitely defined: after a system start, when an unexpected sensor signal is received and when the expected sensor signal has not come after a certain time (train disappeared?). These situations are considered unsafe and can be resolved only by a manual control: if the situation is cleared, the operator may reset the system.

There is seldom a "best" solution for a control system. The solution depends on decisions taken by a system project leader (which state machines to use) as well as on details of the specification of the state machines themselves. Eventually, the specifics of the implementation tools may also influence the solution. When specifying the state machine a designer may minimize the number of states by using Input Actions intensively (Mealy model). Or the designer may not bother about the number of states, trying to achieve a clear, understandable state machine (Moore model).

The TrafficLight control system will be designed as a modular system that can be expanded for any number of tracks, each having three sensors: L, M, R.

"Obvious" Solution

In this case, the designs of single state machines and a system of state machines are strongly coupled, especially in that a hierarchical system will not do here. Let us consider first only the problem that seems to be the basic one: how to identify the train position that determines the traffic lights.

If we think a while about it, we may very soon find an obvious solution: the system has to "know" that a train entered the sensor zone, i.e., has passed the sensor L or R and moves toward the sensor M. This information seems to be sufficient to switch on the red light. When the train is over the sensor M, the red light stays switched on. When eventually the train leaves the sensor M, the red light can be switched off. The information — moving toward (plus staying over the sensor M) and left sensor M — seems to be sufficient to control both lights: red and the flashing yellow. Thus, it seems that, using a hardware analogy, one flip-flop should be enough to control the lamp or, in other words, two states will do.

This solution has one major limitation: it uses the signal edges for control: in hardware it would mean that the rising edge of the L sensor sets the flip-flop; the falling edge of the M sensor resets it. This kind of control is not always possible and is considered unreliable. Anyway, for our exercise, as we want first to show a simple solution we are generous and accept for a while this "edge" based solution and assume that we are able to "detect" the direction of signal changes.

Before we show the error in the solution, we underline once more the difference between the *sunny-day-scenario* and the real world. The simple analysis above has been limited to the *sunny-day-scenario,* which considers only the correct, i.e., the most probable sequence of events (sensor changes according to train movement). If we limit our consideration to the *sunny-day-scenario,* we forget the true control problem. In the discussed example, there are at least the following situations that require consideration, namely, the system behavior:

■ On startup
■ When the "train gets lost" (it entered the sensor zone but never left it)
■ When an unexpected sensor signal occurs, e.g., a sensor M signals the presence of a train though there has been no train yet detected in the sensor zone

If we take into account all what we have said up to now we could specify the state machine shown in Figure 9.22. The diagram shows the basic two states *NoTrain* and *Present* to realize the basic control. On the state transition diagram we use the name Sensor1 for denoting either sensor L if the train comes from the left or sensor R if the train comes from the right. The name Sensor2 means a sensor signaling that the train leaves the controlled zone. The name SensorM means of course the sensor in the middle.

In addition, to make the control system more realistic, we introduced:

■ A state *Start,* which allows the state machine to switch on the traffic lights after the start-up
■ States *Missing* and *Unexpected* to handle the two above-mentioned erroneous situations

If we try to define transition conditions and actions for the *Unexpected* state, we encounter some difficulties. Let us consider the following situation: a short train has entered the sensor zone, the state machine has

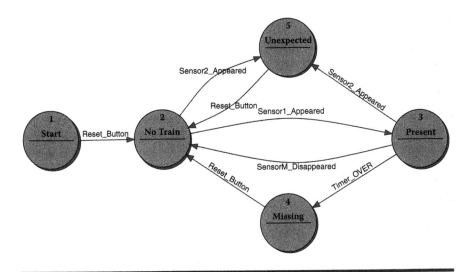

Figure 9.22   Light: the "obvious" but erroneous state transition diagram.

detected the change of the L sensor and has changed its state from *NoTrain* to *Present* where it has switched on the red light and has stopped the yellow flashing light. The train continues its movement toward the sensor M. Eventually, the train leaves the sensor L but it has not arrived at sensor M yet. At this moment the sensor L signals a (new?) train. What should the state machine do now? Our first reaction — change to the state *Unexpected* — will not work. This has been the condition to make the transition from *NoTrain* to *Present*. So, if we use the same condition for the transition from state *Present* to *Unexpected* this transition will be performed immediately when entering the state *Present*. We see that the state machine cannot express the different situations in the analyzed control system with four states. To specify the behavior of the light control we need more states.

The Ultimate Control

After the first unsuccessful trials we jump over the many intermediate solutions that have been explored and present the complete solution.

The control system consists of three state machines: TrafficLight, Light, and Flash. Light is the state machine that realizes the control sequence for one direction. Flash is a state machine that generates the yellow (flashing) traffic light. TrafficLight is the main state machine that controls directly the red traffic light. TrafficLight enables/disables the functioning of Flash by commands `Cmd_Enable` and `Cmd_Disable`.

Light

The state machine Light "follows" the train and at any time presents the train position by its states. We need two state machines Light for one rail: one state machine for each direction. The state transition diagram in Figure 9.23 for Light is as follows: The state transition diagram displays only the states and transition conditions. The complete specification requires the Entry, Exit, and Input Actions, which are only indicated in the diagram by letters E:, X:, I:. The full specification is provided in state transition tables.

If there is no train between sensors L and R, the state machine stays in the state *NoTrain*. Depending on the train length many state sequences are imaginable, e.g.,

■ For a short train, which covers only one sensor at a time

NoTrain -> Coming -> Approaching -> Present ->
Leaving -> Going -> NoTrain

■ For a long train, which covers two sensors at a time

NoTrain -> Coming -> ApprPresent -> Present ->
LeavPresent -> Going -> NoTrain

■ For a very long train, which covers all three sensors at a time

NoTrain -> Coming -> Approaching -> AllPresent ->
Leaving -> Going -> NoTrain

The functioning of the two state machines Light for a rail is mutually exclusive: if one state machine goes into the state *Coming*, the "partner" state machine goes into the state *Disabled*. This arrangement avoids erroneous signaling of failures: a proper sequence of events (sensors) for one direction would be a failure for the other direction.

Comparing the state transition diagram with the diagram specified in the *Design example — Traffic light control* section in Chapter 5, we find the states used there in the main loop:

NoTrain -> Coming -> Approaching ->
Present -> Leaving -> Going

of the present state machine. This main loop has been extended by three states: *ApprPresent, AllPresent,* and *LeavPresent,* which are needed as we allow trains with any length. The other additional states have been introduced to cover erroneous situations resulting from false signals from

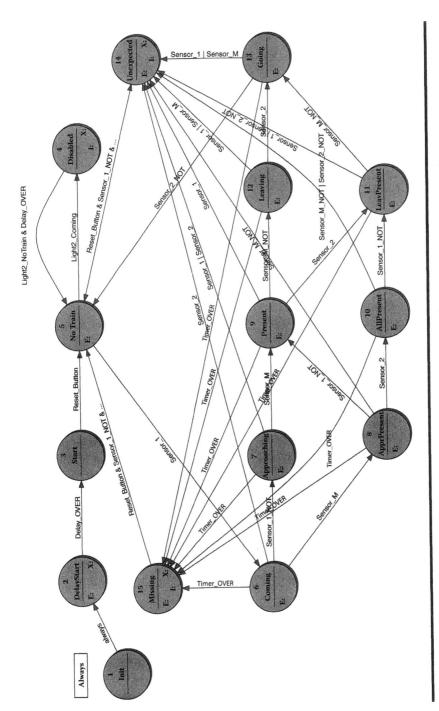

Figure 9.23   Light: the state transition diagram.

Figure 9.24 TrafficLight: a combinational system will do.

sensors. Obviously, the hardware solution was a rather simple example covering only the *sunny-day-scenario*.

TrafficLight

Because the state machine Light represents, at any time, the train position, the state machine TrafficLight can be a simple combinational system. In other words, it is just a degenerated state machine with one state *Init* (Figure 9.24).

The behavior of the two traffic lights — red (controlled by TrafficLight) and yellow (controlled by Flash) — is then defined by states of the state machine Light in Table 9.1. We mention in passing that the state machine Light specified in that project is a kind of Parser type (see the *Application based state machine models* section in Chapter 6) state machine: it translates (by parsing the sensors L, M, R) the train position into states. The states are then used by the state machine TrafficLight to decode the situation using a pure combinational system for controlling the crossing lights.

To simplify the logic equations the states are combined in complex Conditions, e.g., the condition `TrainComing` is an OR combination of six states (*Coming, Approaching, ApprPresent, Present, AllPresent, Leave-Present*) and describes situations when the train is between Sensor1 and Sensor2 moving toward M or is already on M. Using the complex Conditions the following equations specify the On/Off signals for the Red light:

```
RedLight_On = Start | TrainComing | Error
RedLight_Off = Disabled | NoTrain | TrainGoing
```

Similar equations can be defined for the Yellow (Flash) light:

```
Flash_Enabled = Start | Disabled | NoTrain |
                TrainGoing | Error
Flash_Disabled = TrainComing
```

For several rails the OR-combination for On signal and AND-combination for Off signal will do. For instance, for two-rail railway we get the following equations for the Red light:

Table 9.1 TrafficLight: Dependencies between Light States and Red and Yellow Lights

Light state	Condition	TrafficLight (red) Off	On	Flash (yellow) Disable	Enable
Start	Start		X		X
Disabled	Disabled	X			X
NoTrain	NoTrain	X			X
Coming			X	X	
Approaching			X	X	
ApprPresent	TrainComing		X	X	
Present			X	X	
AllPresent			X	X	
LeavePresent			X	X	
Leaving	TrainGoing	X			X
Going		X			X
Unexpected	Error		X		X
Missing			X		X

```
RedLight_On  = Start1  | TrainComing1 | Error1 |
               Start2  | TrainComing2 | Error2
RedLight_Off = (Disabled1 | NoTrain1 | TrainGoing1) &
               (Disabled2 | NoTrain2 | TrainGoing2)
```
and for the Flash light:
```
Flash_Enabled  = (Start1 | Disabled1 | NoTrain1 |
                 TrainGoing1 | Error1) &
                 (Start2 | Disabled2 | NoTrain2 |
                 TrainGoing2 | Error2)
Flash_Disabled = TrainComing1 & TrainComing2
```

Flash

The yellow light is not controlled directly by the state machine TrafficLight. There is a simple state machine Flash (see Figure 9.25), which generates the flash-cycle and controls the yellow light. The state machine TrafficLight controls the behavior of the state machine Flash using commands: CmdEnable and CmdDisable.

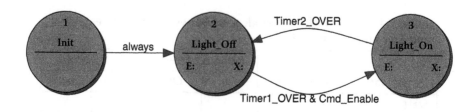

Figure 9.25   Flash: the state transition diagram.

System for Two-Track Railway

The diagram in Figure 9.26 shows the state machines system required for a two-track railway. It is interesting to note the unconventional system structure. It is not a hierarchical system. The main state machine TrafficLight, which in fact controls the traffic light, is a combinational system that uses the states of the state machines Light as inputs.

The state machines Light create a layer that "translates" the train movements into definite train positions. Knowing the train positions, the control problem simplifies to a pure combinational system solved by TrafficLight. The approach can be extended to more tracks, and we may

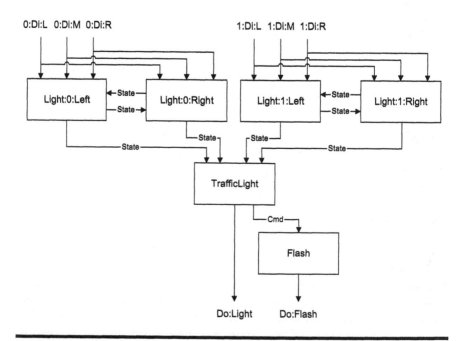

Figure 9.26   State machines for controlling a two-track railway.

notice that, once a state machine has been designed, further instances can be added to the project very easily.

The implementation of the TrafficLight project in StateWORKS is shown in Appendix N *Traffic light project*. You will see that there is nothing much in the *Init* state transition table of the state machine TrafficLight, and that the logic is located in the *Always* table.

Summary

Is the solution the best one? Nobody can answer this question. We like it and it works perfectly, so we stopped searching for better solutions.

We may have questions, especially considering the state machine Light. It may seem an overkill to use so many states. Considering the basic representation of train movement, we started with two states (*NoTrains* and *Present*) and ended with nine states (*NoTrain, Coming,... Going*). We have shown that the minimal solution could not work but we have not proved that we really need nine states. Maybe a smaller number will do. We have of course tried other solutions. If we were to design a hardware system with a limited number of flip-flops (nowadays this does not seem to be a problem either), we would try to minimize the number of states. Using Automata Theory design methods we can arrive at a state transition diagram with a minimal number of states (see the minimal solution with six states in Figure 5.3, *Design example — Traffic light control* section Chapter 5).

But the solution is not adequate as the software solution covers more than the requirements defined for the hardware project. But we do not want to invest more time tuning the number of states as there are two problems with Automata Theory approach:

■ How many people know and are able to effectively use the Automata Theory methods?
■ A solution with the minimal number of states is often less comprehensible than a system with intuitively defined states.

In a software implementation the number of states does not significantly influence the needed memory, so other factors are important. In software implementations we should choose clear, understandable solutions. This should not be understood as encouragement to overuse states. In this specific control system we have the impression that the states used represent very well the train position, which is the essential information. Paraphrasing a well-known rule we would recommend:

We use as many states as we find necessary but no more.

With this example of traffic light control we would like to send a message: do not underestimate the control problem. At a first glance, the traffic light control may appear a trivial exercise. It may be simple if we limit our considerations to a *sunny-day-scenario* with one train going in one direction, but such a solution is without any practical value. It is a non-trivial problem if we discuss it seriously, taking into account the real situations that may occur in such a system. The task of a control system is to react not only to expected, routine sequences but also to manage unexpected situations that may appear once a year or even once in the system lifetime. Correct actions in these rare situations may save human life or save costly damage to hardware equipment.

The example demonstrates how to cover unusual situations in a control system. Problems arise from the rarely occurring but dangerous, erroneous situations that make the design of control systems difficult and justify all the effort needed to design a good control system.

One must bear in mind that the complexity of a state machine is not just a matter of the numbers of states and transitions. We have encountered situations where a very complex process, in fact extremely difficult to express correctly in code, has been expressed by means of a state machine with only four or five states. This state machine seems to be quite simple, until one examines the very complex expressions governing state transitions, and the lists of entry and exit actions associated with some of the states. Then one realizes that one is dealing with a complex entity and begins to understand why coding would be difficult, and where the advantage of a state machine approach lies, with its coherent framework for expressing the complete behavior of the system.

We should also not forget that examples we can illustrate in a book cannot be too complex to be easily appreciated by the reader. The entire control problem will be seen when designing a true traffic light control system, being confronted with complete specifications considering available sensors, actuators, and especially in this case safety requirements.

Conclusions

The examples we gave represent fragments of a typical design procedure. We start to analyze the control system and produce the first trial solution. Doing this we learn better the requirements. After several trials we understand the problem and we do the first system design where we decide about the overall structure of the system: inputs, outputs, type and number of state machines and their interface (in fact, we decompose at this moment the control task into manageable entities and determine how these entities cooperate). The system design may not be the ultimate one. It may be changed if during the preparation of state machine specifications we encounter problems whose solutions require changes to the system design.

We could not show all the steps, errors made, all the defeats and triumphs that accompany the design process. We presented just the first unsatisfactory trial and then the ultimate solution, ultimate from our point of view. We would not be surprised if somebody finds a better solution.

Implementation

A state machine or a system of state machines specified according to rules presented in Part II described fully the behavior of the controlled application. The specification contains all details; i.e., there is no need to complete it later with some features which were (intentionally?) left for the implementation. Even if we really forget something we can always return to the specification and add the missing feature as we have shown that there are no limitations in the presented method.

If the specification is complete we could think about its execution, at least for testing purposes. Having an executor that carries out the specification we would get a possibility to test it for logical errors. After those tests we are sure that the specification is nearly perfect. If we make still another step and add to the executor of the specification input/output and user interfaces there is no need to strive after another (coded) implementation — that executor would be the implementation.

As we have gained experience designing the examples we can now expand the already defined design rules formulating the following recommendations.

Designing a State Machine

- Choose good expressive names for states and conditions.
- States of a state machine must contain the full control information on its abstraction level.
- Start with *sunny-day-scenarios* and refine.
- The highest priority should be given to transitions to error signaling states; the expected transition should be placed at the end in the state transition table, with lowest priority.
- Use timers guarding against deadlocks in state machines that communicate with external signals.
- Consider the integration into a system of state machines.

Designing a System of State Machines

- Use a hierarchical structure if there are no obvious alternatives.
- Use Master–Slave (command–state) interfaces for communication between state machines.
- Keep strict handshaking rules for synchronizing Master with Slaves.
- Use bottom-up design.
- Avoid corruption of the hierarchy by "wild" links among state machines.
- Use timers guarding against deadlocks only in state machines that communicate with external signals.

We see that by following the rules we are able to achieve an executable specification, and the executor that executes the specification can be the application or at least a frame for the application. This kind of software realization is the topic of the following part.

Part III

STATEWORKS: PRINCIPLES AND PRACTICE

Chapter 10

StateWORKS

Virtual Environment and Vfsm

In the *Any class of signal may "contain" the control value* section in Chapter 6, we have shown that instead of trying to express all control expressions in terms of Boolean values we may operate on names describing the control inputs. Formalizing this approach and completing it with positive logic algebra gave rise to the **Vfsm method**. Vfsm stands for Virtual Finite State Machine, where the name "virtual" characterizes the environment in which the Finite State Machine operates. The concept of the Vfsm method was first presented by Wagner.[1,2] The achieved results were shown by Wagner et al.[3-5]

The basis of the Vfsm *method* is the concept of a **virtual environment,** which is defined by three **sets of names**:

- **Input names** representing control values
- **Output names** representing actions
- **State names**

The Input and Output names represent pure control information derived from some real signals. To distinguish them from real signal values and to underline their abstract content, we consider and call them as virtual values. State names do not have any equivalents in the real world — they are a product of our imagination and help us to describe behavior. As such, there is no need to call them *virtual* but state names are by definition abstract and therefore fit perfectly into the virtual environment.

The virtual environment implements the idea of separation between data and logic operations. The virtual environment does not know the data — it is a pure control concept.

The StateWORKS Development Environment

StateWORKS is the practical implementation of ideas formulated in Part II of this book. Introducing control values and actions in a virtual environment should produce a software development system in which the application behavior is specified and not coded. In other words we are discussing an executable specification.

StateWORKS uses a state machine model for expressing behavior. The used state machine model must be defined to the last details because the results of the specification are not vague descriptions of intention but are used by the runtime system. Therefore, StateWORKS defines a state machine execution model, which a designer must have in view while designing a state machine.

A state machine is a decision machine. Decisions are logical expressions governed by rules of Boolean algebra. Unfortunately, we cannot apply the rules of Boolean algebra directly to Input names, which are non-Boolean values. Therefore, StateWORKS introduces a specific *Positive logic algebra,* which is a set of rules for using logical operators in a virtual environment.

Positive Logic Algebra

Sets of names that define the virtual environment are used to build *virtual conditions, virtual input,* and *virtual output.*

The **virtual conditions** are a set of names — in fact control values. The *virtual conditions* are used to express Transition and Input Action conditions:

- A *virtual condition* defines a (Boolean) AND operation on names in the set.
- A group of *virtual conditions* defines a (Boolean) OR operation on *virtual conditions.*

For instance, using conventional set notation:

`{Di_ON, Temp_HIGH} {Timer_OVER}`

specifies two conditions. The first one reads: Di is ON AND Temp is HIGH.

The second one reads: Timer is OVER. Together they specify a logical expression:

```
(Di_ON & Temp_HIGH) | Timer_OVER
```

To hide from the user the set definitions and operations, which are less comprehensible than Boolean equations, we use in StateWORKS Studio the logical expression with the symbols **&** for the logical AND and **|** for the logical OR operation.

With this arrangement we achieve a user-friendly development environment where logical conditions are expressed in the way to which users are accustomed. As the NOT operation is not applicable here and is not used, we call the rules **positive logic algebra**.

The **virtual input** (**VI**) is a set of names — control values describing the actual input (subset of Input names). For instance:

```
{Di_ON, Temp_LOW, Timer_OVER}
```

specifies the input well described by the names used: the digital input Di is ON and the Temperature is LOW and the Timer is OVER.

VI contains by default a name **always**. Thus, VI can never be empty; it contains at least the name always. The existence of that name simplifies specification — the name is used to specify a condition that is always true. Actually the previous example of VI has the strict form:

```
{always, Di_ON, Temp_LOW, Timer_OVER}
```

The **virtual output** is a set of names — actions that are needed for output definitions (subset of Output names). For instance:

```
{Do_On, Cmd_Stop}
```

specifies actions to be performed: the digital output Do is On and the command Cmd has the value Stop.

The Vfsm Execution Model

A Vfsm is thus a state machine specified and operating in a virtual environment:

- The Transition and Input Action conditions are expressed using *virtual conditions*.
- Actions are expressed using *virtual output*.
- The actual input is represented by VI.

The Vfsm Executor compares continuously the VI with those Transition and Input Action conditions relevant to the present state and works out the transitions and actions to be done:

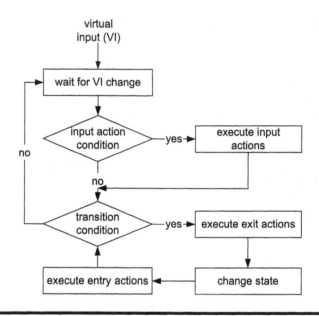

Figure 10.1   The Vfsm execution model.

- Perform an Input Action if the Input Action conditions are fulfilled.
- Change a state if the Transition condition is fulfilled.
- Perform all Exit (in the present state) and Entry (in the next state) Actions if the state changes.

The Transition or Input Action condition is fulfilled if any set of corresponding conditions is a subset of VI.

The complete execution model of the Vfsm is described by the flowchart shown in Figure 10.1. The Executor waits for a VI change. If it occurs, the input conditions of all Input Actions are checked, and if due, the corresponding Input Actions are carried out. Then the Transition conditions are checked — the first transition that is due is carried out: the Exit Actions are executed, the state is changed, and the Entry Actions in the new state are executed. In the new state the Transition conditions are checked, and if due, the next transition is done. This process is continued until there are no transitions due in the state that has been reached — then the Executor returns to the waiting point.

The commitment to this specific execution model is not arbitrary. It is based on practical experience and many years of discussions and considerations. We believe that it is a good compromise among several requirements, when considering the imaginable sequences of transitions and actions. The characteristics of the Executor model can be summarized as follows:

- An event is a change of the virtual input and not a change of object data. In addition, providing the same data value does not generate an event — an event is produced by a change.
- A single event is able to trigger a series of several state transitions. All Entry Actions in entered states (intermediate and final) are performed but only Input Actions due in the starting state are carried out; Input Actions in intermediate states are ignored.
- The first state transition, which is due, is performed — the sequence of checking determines the priority (defined by the sequence in the state transition table).
- All Input Actions that are due in the present state are carried out but their execution sequence is unpredictable. Hence, we must not make any assumption about their order: any required sequence of outputs can be only achieved by arranging a sequence of states.
- The same rule relates to Entry and Exit Actions. If many of them are carried out in a given state, their sequence is unpredictable.

Objects

The Vfsm definitions are sufficient for coded implementations according to rules described in Part II of this book.

The primary goal of StateWORKS has been to completely eliminate coding of the control logic. To achieve this, StateWORKS operates on objects that store values and translate them into control values and sets outputs in response to actions. In other words, an object is an interface between the physical signals and the virtual environment. To manage objects effectively they are organized in a **Real Time Data Base** (RTDB; see also the *StateWORKS execution environment* section below).

The physical signals are very heterogeneous. It is difficult, maybe impossible to cover them with one universal object. Therefore, State-WORKS defines various object types covering a large variety of imaginable signals: it is possible to add extra objects if needed. There are two groups: input and output objects. There is also a group of objects that possess both features: those of inputs as well as output objects. Table 10.1 summarizes the basic control features of objects.

The full characterization of objects is more complex. The table contains only control features and actions required for the virtual specification. The complete descriptions are supplied in the following chapters discussing details of objects, their properties, and use.

Table 10.1 RTDB Objects

Object	Description	Value Type	Control Values	Actions
VFSM	Vfsm	State (integer > 0)	state -> number	—
CMD as CMD-IN	Command	Command (integer > 0)	name -> number	Clear
CMD as CMD-OUT	Command	Command (integer > 0)	—	*name -> number*
TI	Timer	Counter value	RESET, STOP, RUN, OVER, OVERSTOP	Reset, Stop, Start, ResetStart
CNT	Counter	Counter value	RESET, STOP, RUN, OVER, OVERSTOP	Reset, Stop, Start, ResetStart, Inc, Dec
ECNT	Event Counter	Counter value	RESET, STOP, RUN, OVER, OVERSTOP	Reset, Stop, Start, ResetStart, Inc, Dec
UDC	Up/down counter	Counter value	UNDEF, DEF, CHANGED, INIT	Clear, Up, Down
SWIP	Switch-point	Supervised value	OFF, LOW, IN, HIGH, UNDEF	Off, On
STR	String	Supervised string	OFF, INIT, MATCH, NOMATCH, DEF, ERROR	Off, On, Set
XDA	Number	Integer > 0	*number > 0*	*number > 0*
OFUN	Number	Integer	number	number
DI	Digital input	Boolean value	UNKNOWN, LOW, HIGH	—
NI	Numerical input	Number (type defined by a Format property)	UNDEF, DEF, CHANGED	—

Table 10.1 (continued)   RTDB Objects

Object	Description	Value Type	Control Values	Actions
DAT	Any input	Number (type defined by a Format property)	UNDEF, DEF, CHANGED	—
PAR	Parameter	Number (type defined by a Format property)	UNDEF, INIT, DEF, CHANGED	—
DO	Digital output	Boolean value	—	Low, High
NO	Numerical output	Number (type defined by a Format property)	—	Off, On, Set
AL	Alarm	Number	—	Coming, Going, Staying
TAB	Demulti-plexer	Integer > 0	—	*Number (index)*
UNIT	List of objects	—	—	—

Objects representing pure input features are DI, NI, DAT, and PAR. They store the input values (or parameters) and generate the control values.

Objects representing pure output features are: DO, NO, AL, and TAB. They are controlled by actions and they set the output values.

The largest group contains objects that are both inputs and outputs: CMD, TI, CNT, ECNT, UDC, SWIP, STR, XDA, and OFUN. The last object — OFUN — is a kind of software interface allowing calls of user-written, application-specific functions which extend the range of possible object types.

VFSM is an object that stores the behavior specification of the state machine, the virtual input, and a state (control value).

UNIT is of no direct interest in relation to control. UNIT is just a list of objects that are used in the physical grouping of I/O handlers and Output functions.

All object types have also **Description** properties: **Text** and **Link**. Those properties are outside the scope of the book. They are used for XML documents generated by StateWORKS Studio and are described in StateWORKS Studio help.

State Machine Defines Object Control Values

All input or I/O objects possess control features; i.e., they generate control values used for state machine specification. Those values reflect the states of input signals (e.g., DI, NI), data (e.g., PAR or DAT), or a state of a computer resource used (e.g., TI, CNT). In other words, control values present aspects of the behavior of those objects. That behavior can in fact be described by a state machine. We use this kind of description in the following chapters, which specify the RTDB objects. For each object type we draw a state machine illustrating its behavior (see, e.g., Figure 11.1 for DI, Figure 12.1 for DAT, Figure 14.1 for CNT, etc.). States (written in uppercase) of those state machines are the control values of the corresponding objects.

Signal Lifetime

Objects store data and corresponding control values. Both data and control values may lose their meaning in certain situations. In general, we may speak about (limited) lifetime of control signals. This problem relates especially to data coming from outside: peripheral devices (hardware) or files.

Hardware itself or the path from hardware to the RTDB may break down and the information delivered by drivers and I/O handlers ceases. Especially objects of type DI and NI can be affected. Therefore, these object types should have a control value (UNKNOWN *or* UNDEF), which informs directly about the arisen situation.

A similar problem may occur when loading parameters from a disk or network on system start-up. For instance, a missing file should be signaled with a control value (UNDEF) of the PAR object.

A special case is the use of CMD objects. A command is a signal whose value can be only replaced by another command. A command is not generated by some physical events — a command is issued as a result of a designer's decision. In other words, a designer defines the lifetime of a command. We have discussed the problem already (see the *Signal lifetime* section in Chapter 7).

The situations caused by hardware malfunctions could, of course, be solved using separate signals, but integrating detection and signaling of such erroneous cases works more smoothly if the control value of affected data provides a corresponding value.

We encounter a similar problem when writing functions managed by OFUN objects. When writing such output functions we have to provide a return value that informs about erroneous situations — this value can then be treated as a control value, telling us about any unfortunate calculations.

Behavior Specification

The task of a VFSM is to react to input changes, working out appropriate outputs. Input stimuli are stored in the RTDB and are "seen" by VFSM as control values. Control values are static signals that reflect the outside world and change according to changes of physical input signals. The outputs are set by VFSM as actions, which are then transformed into physical outputs. Having this model in mind we specify the behavior using the state machine execution model shown in Figure 10.1. We may do it in a development environment such as StateWORKS Studio, which makes several editors available to create **State Transition** diagrams (**ST diagram**), **State Transition** tables (**ST** table), **State Machines System** (**SMS**) diagrams, and Objects and their Properties. The ST diagram and the ST tables are used to specify a single state machine. The SMS diagram helps in the specification of a system of multiple state machines.

First, we must discuss the use of Input and Output names. Any object has control values that may be used for control specification. As we have several objects of the same type we must distinguish among their control values. We do it by naming them according to their meaning. For instance, a control value `OVER` for a timer object Timeout would get the name `Timeout_ELAPSED` and for another timer object — `MyPatience_EXPIRED`. The same applies for actions. For instance, a Do High value switching the Motor will get the name `Motor_On` and another Do High value closing a valve will get the name `Valve_Close`.

The necessity of inventing State names is obvious — those names are the basis of any specification that uses state machine models.

StateWORKS Studio uses diagrams and tables already introduced and discussed in Part II of this book. For now, we confine ourselves to an example of a table explaining the detailed specification. For this purpose we take a table from the previously discussed state machine Pressure (see Figure 10.2). The table specifies a state *Starting* using Input names, Output names, and State names.

The Input names that we need to create are the names assigned to those control values we shall use, taken from the set of possible control values of the objects we wish to use. This means that before we create

Several activities are initiated. Waiting for Pressure acknowledgments. Due to a Timer missing acknowledgment leads to return to the Idle state. Too hot a pump leads to the PumpError state. Both erroneous situations generate corresponding alarms. Error by accessing the output function returns the state machine also to the state Idle: it does not make sense to supervise the pressure without having proper pressure limits (corresponding alarms are generated in Always table).

Starting	Entry action	MyCmd_Clear SetPressure_Set Counter_ResetStart Timer_ResetStart Ofun_CalcLimit
	eXit action	Timer_Stop
	RequiredPress_CHANGED	Timer_ResetStart
	Timer_OVER	Al_PressureError
PumpError	Pump_TooHot	
Idle	Timer_OVER | Ofun_OwnerError | Ofun_ParameterError	
Regulating	Press_OK	

Figure 10.2   An example of an ST table.

names we have to decide which objects will be used by the state machine. Then we create names for object values. For instance, because we have to supervise the pressure we need the SWIP object. The SWIP object (see the *Getting the control value (SWIP)* section in Chapter 12) has five possible control values: OFF, LOW, IN, HIGH, and UNDEF. So, we shall create a Press_OK name for the control value IN.

Similarly, the Output names used signify actions and are created using possible actions (commands) on the output objects. For instance, as we need a timer (TI object type, see the *A timer (TI)* section in Chapter 14) to guard the state *Starting* we invent an Output name Timer_ResetStart corresponding to the command ResetStart for a TI. Note that a timer is controlled by actions, but it also generates Input names (e.g., Timer_OVER created on the control value OVER). Other Input names (Pump_TooHot and RequiredPress_CHANGED) have been created in a similar manner on input objects: Di and Par. Similarly, other Output names (MyCmd_Clear, SetPressure_Set, Counter_ResetStart, Ofun_CalcLimit, Al_PressureError) have been created on objects: MyCmd, No, Ecnt, Ofun, and Al.

Specification of state independent actions. If the Required Pressure value changes:
- the Error Counter is reset,
- the Pressure Limits are recalculated,
- the Pressure value is set.

Always	RequiredPress_CHANGED	Counter_ResetStart Ofun_CalcLimit SetPressure_Set

Figure 10.3   An example of a table *Always*.

The State names are just freely invented names that we need to better describe our Vfsm design to help understand it later.

There are three possible transitions from the state *Starting* to the states: *PumpError, Idle,* and *Regulating*. The positions of those transitions in the table are important — they determine priorities, with *PumpError* having the highest priority. In contrast to transitions the actions are not prioritized — their positions in the table are insignificant. This rule applies to all types: Entry Actions, Exit Actions, and Input Action expressions.

The Input Actions specified in the State transition table may not be complete. If the state machine has actions specified in the table **Always**, these are equivalent to Input Actions specified in each State Transition table. Such Input Actions are placed in the table *Always* as a shortcut. They contain state-independent Input Actions, i.e., actions that will be checked and carried out if due in any state. So, in the example the Input Actions in the state *Starting* are complemented by the Input Actions specified in the table *Always* (see Figure 10.3).

One state machine has limits. Those are not limits of the state machine itself, but rather of the human beings who are able to comprehend and keep under control problems of limited size. Therefore, we should partition a more complex control problem into several state machines. The state machines communicate among themselves as described in Chapter 15 *VFSM and Its Interfaces*.

To illustrate the communication using Commands and State we take another already known example — the system of state machines in the project Pumps. We will analyze the state *StartingPressure* (see Figure 10.4). Main is a Master of three state machines: two machines of type Pressure and one of type Device.

On starting a specification of a system of state machine we have a problem. We have already specified state machines as regards their behavior, but each may have many instances; effectively we have specified behavior of certain types of state machines. As the system does not yet exist we have to assume that it will be created at some stage and for now

An intermediate state where the state machine waits for Pressure slaves acknowledgments to the commands Start sent to them on entering the state.
Receiving positive acknowledgments: Starting or Regulating it goes to the state.
PressureStarted, otherwise it returns to the state Idle. Receiving Error states from slaves it acknowledges that sending the command Break to the slaves.
Note missing timer - the slaves are responsible for that, the Main master control is based exclusively on slaves' states.

StartingPressure	Entry action	Pressure1Cmd_Start Pressure2Cmd_Start
	eXit action	
	Pressure1_PumpError	Pressure1Cmd_Break
	Pressure2_PumpError	Pressure2Cmd_Break
PressureStarted	(Pressure1_Starting | Pressure1_Regulating) & (Pressure2_Starting | Pressure2_Regulating)	
Idle	Cmd_Stop | Pressure1_PumpError | Pressure2_PumpError	

Figure 10.4   An example of a state in a Master state machine.

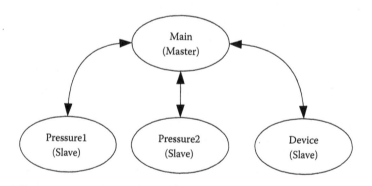

Figure 10.5   The idea of Master–Slave system in behavior specification.

we just declare how many Slaves of a given type any Master will have. Thus, we prepare for some Master–Slave links. For the discussed example Pumps, we think about something like in Figure 10.5.

From the SMS diagram shown later in Figure 10.6 we learn that the Main state machine "sends" commands (actions) to its Slaves, Pressure1, Pressure2, and Device, and uses the states of its Slaves as control values.

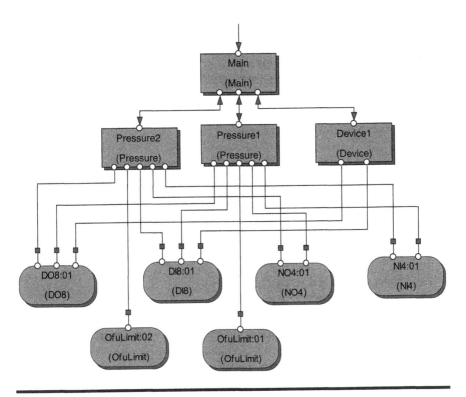

Figure 10.6   An example of an SMS diagram.

In the state transition table of the state *StartingPressure* the details of the control are seen. To specify the state we have decided that the state machine Main will have two equal state machines of type Pressure as Slaves and one of type Device. The states of state machine Device are not used in the discussed state *StartingPressure*; hence, we concentrate on the two Pressure Slaves. Hence, we have declared two VFSM objects (see *A virtual finite state machine (VFSM)* in Chapter 15) of type Pressure and two CMD objects (see *A command (CMD)* in Chapter 15) of type Pressure in the Main VFSM. By declaring those objects the states of the Pressure Slaves are made available in the Main specification and can be used as control values. Thus, we create the names:

- `Pressure1_Starting` (on the State name *Starting* of Pressure1)
- `Pressure1_Regulating` (on the State name *Regulating* of Pressure1)
- `Pressure1_PumpError` (on the State name *PumpError* of Pressure1)

We create similar names on State names of Pressure2:

- `Pressure2_Starting`
- `Pressure2_Regulating`
- `Pressure2_PumpError`

Similarly, we create Output names:

- `Pressure1Cmd_Start` (on the command name `Cmd_Start` of Pressure1)
- `Pressure1Cmd_Break` (on the command name `Cmd_Break` of Pressure1)

and Output names on command names of Pressure2:

- `Pressure2Cmd_Start`
- `Pressure2Cmd_Break`

Having those names we use them to specify the state machine:

- Input names for declaring state transition and Input Action conditions
- Output names for declaring actions

We see that the usage of VFSM states as control values corresponds to creating Input names by using any other RTDB objects (this will be even more obvious if we understand that object control values are states of state machines that describe their behavior).

The use of Slave commands as actions requires further discussion, which will be done in detail in *A command (CMD)* in Chapter 15. At this point we would only outline the topic, referring to Figure 15.3 in Chapter 15. A command is stored in an object of type CMD. The CMD object has two aspects. On the one hand, it belongs to a certain state machine where its values may be used as control values (Input names). On the other hand, the commands must be set by somebody. In the example, CMD objects, which belong to state machines Pressure1 and Pressure2 (as well Device), are also accessed by the state machine Main. The state machine Main uses actions such as `Pressure1Cmd_Start` or `Pressure2Cmd_Start` to send corresponding commands to Slaves: Pressure1 and Pressure2. The same CMD objects are treated in Pressure1 and Pressure2 state machines as input commands and used in Transition and Input Actions condition to control their behavior.

System Specification

All that we have said so far refers to a (virtual) specification of state machines. Behavior of each state machine is specified in ST tables supported by an ST diagram. It is similar to a definition of a type or a class in a programming language.

The next step is to specify objects needed to implement the designed control system, which is similar to declaring variables, e.g., instances of classes. This specification includes definitions of links between objects:

- Master → Slaves
- Supervised objects (e.g. NI) → Supervising objects (SWIP, STR)
- Inputs and Outputs (e.g. DI) → I/O handlers
- Used objects → Output functions (OFUN)
- Parameters (PAR) → Parameterized objects (e.g. TI)

In this specification process we determine also object properties (see descriptions of objects in the following chapters).

Having specified the system we are able to get the entire view of the system of state machine in the form represented by an example in Figure 10.6. The SMS diagram shows all state machines and UNITs used and how they are linked. The lines with arrows display the Master ↔ Slave (command ↔ state) links: an arrow to a Master (the upper state machine) indicates the Slave's state; an arrow to the Slave (the lower state machine) indicates the Master's command. Other bindings (e.g., XDA) or to UNIT and OFUN are indicated using lines with a small rectangle. By placing the cursor over an arrow or a rectangle a tool tip appears displaying name(s) of involved objects.

Now we are able to build the system, producing the files required by the StateWORKS execution environment.

The StateWORKS Execution Environment

The core of the StateWORKS concept is the executable specification. The results of the state machine specifications are files that contain the entire knowledge about the application's behavior. The form of the files is adjusted to the RTDB; i.e., if loaded into the database they become executable. The RTDB is the heart of a StateWORKS runtime system. The full application requires output functions and I/O handlers. Output functions contain that application specific functionality, which cannot be realized by RTDB objects. The I/O handlers are the interfaces to peripheral devices — to the controlled system. For completeness we should also mention the user interface, but it is not an integral part of the StateWORKS runtime system — it is a foreign program, which communicates with the RTDB via TCP/IP.

RTDB-Based Runtime System

We have already introduced the term RTDB as a database containing objects that represent all the signals required for control of an application. But the RTDB is much more than that. It is not only a collection of objects — it contains the Vfsm Executor, which carries out control according to the Vfsm execution model (see Figure 10.1).

The RTDB is available as a library and is used to build applications. Figure 10.7 shows the basic components of such an application. It contains:

- The RTDB with the Vfsm Executor
- The I/O handlers
- The Output functions

The RTDB with the Vfsm Executor is a fixed code independent of the application. The specifics of the application are determined by the Control

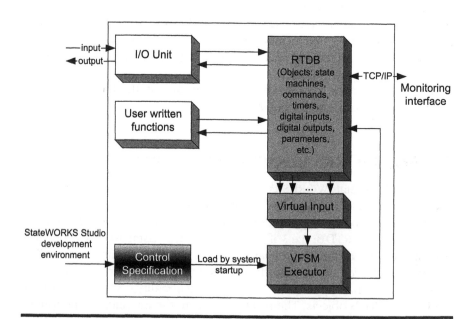

Figure 10.7   A Vfsm runtime system.

Specification, the I/O handlers, and the User-written output functions. The Control Specification consists of a set of files produced by the StateWORKS Studio and read at start-up of the runtime system.

The runtime system communicates with the controlled system via I/O handlers, which deliver input signals to the RTDB and pass output signals from the RTDB. RTDB objects are accessible using a TCP/IP link — this is mostly used for user interfaces, but it may be also used as a substitute for I/O handlers, or for various other purposes.

The Vfsm runtime system is an event-driven system, where the event is a change of a control value. A data change in the RTDB (input via I/O handlers, parameter via user interface, internal event like timer expiration) may cause a change of control values, i.e., a change of state machine virtual inputs (one or more). Such a change triggers the Vfsm Executor, which performs actions and state transitions according to information in Control Specification. Those state transitions and actions may change other control values and in turn corresponding virtual inputs; all of them are treated one by one in the emerging sequence until all events are fully processed.

Output Function

The applications will often require data processing that cannot be performed by RTDB objects. Typical examples are calculations of limits, time, etc. The RTDB contains an object OFUN (see *An interface to a user written function (OFUN)* in Chapter 15) to handle such problems. The OFUN object is a software interface allowing expansion of the RTDB functionality. This object itself cannot process data but it manages the actions which call the required C/C++ function(s), and it serves as a source of control values that may be used for behavior specification. The OFUN control values are the return values (integers) of the called functions. The called function may return calculation results; at least the information whether the calculations were successful or failed — all of those return values represent control values.

The definition of an OFUN object is based on a UNIT, which defines RTDB objects accessible by the called function. As a simple example, the Pressure state machine uses a SWIP object to supervise a pressure represented by numerical data stored in an NI object. The allowable pressure range is defined by Limits of the SWIP object. In this example the Limits cannot be constant values as they depend on the required pressure value and the allowed pressure deviation from it. All those dependencies are shown in Figure 10.8. The OFUN calls a function that calculates the Low Limit and High Limit for the SWIP using the parameter values: Required Pressure Value and Pressure Deviation.

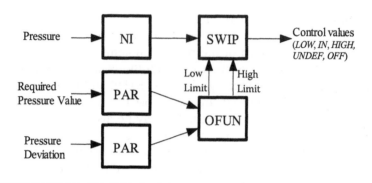

Figure 10.8   Example of dependencies in OFUN definition.

Thus, the UNIT used by the OFUN must contain a list of objects which the calling functions are to access:

■ PAR: Required Pressure Value
■ PAR: Pressure Deviation
■ SWIP: Supervising Pressure (NI)

The application contains two Pressure state machines. Therefore, the system has two OFUN objects and each has its own UNIT (OfuLimit:1 and OfuLimit:2 in Figure 10.6, see also Appendix M *Output function CalcLimit()*). A VFSM object may be used as a substitute of a UNIT if it contains all objects required by the output functions.

I/O Handler

The RTDB possesses a software interface for linking RTDB objects with the outside world (Inputs and Outputs). The interface is a set of methods of C++ classes, which are used to write I/O Handlers. The use of I/O Handlers is not limited to the obvious communication with hardware peripheral devices. From a programmer's point of view an I/O Handler (and its definition UNIT) is a gateway to the RTDB. We may access the RTDB objects for any (reasonable) purpose. For instance, instead of using Output functions defined explicitly for such purposes (via OFUN), we may concentrate all the application-specific programming part in I/O Handlers.

Similar to output functions, when programming an I/O Handler the software developer needs a list of RTDB objects that must be accessed from the I/O handler. The list is defined by a suitable UNIT. The obvious candidates for the list are DI, DO, NI, NO objects, which represent digital and numerical inputs and outputs. Of course, any other object may be declared in the list if required. As for OFUN, a VFSM object may be used as a substitute for a UNIT if it contains all objects required by the I/O Handler.

User Interface

The RTDB contains the TCP/IP interface for communication with objects. The RTDB is a server, and foreign programs that want to communicate with the RTDB must be clients. The server–client model of the StateWORKS runtime system is shown in Figure 10.9 and uses a port with two sockets: Request/Reply and Event/Ack. If connected, a client may send a Request (or Poke, Advise), which is answered with a Reply. If "advised" (when registered as a client*), the client receives events that have to be acknowledged.

RTDB objects are complex items. Each object contains several attributes (values), which can be separately read or changed. Two examples are shown below (Table 10.1 and Appendix S: *Attributes of RTDB Objects* show the attributes of all objects):

- DI: Peripheral Value (PeV), Service Mode (SvM), Service Value (SvV), Trace (Trc), and State (Val)
- TI: Time-out value (CnC), Running counter value(CnR), State name (StN), Trace (Trc), Timer base (Uni), and State (Val)

A client may request or change (if it makes sense and is allowed) any object attribute. A client may register to any object attribute so as to be notified of any change. Hence the RTDB server–client model incorporates both the synchronous transmission for Request and the asynchronous transmission for Events sent as notification to registered clients for those attributes.

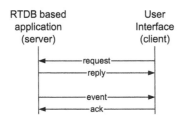

Figure 10.9   The RTDB TCP/IP Server–Client communication.

---

* Not in a TCP/IP sense but as an object user.

For implementation of the TCP/IP communication we can use a static or dynamic (DLL under Windows) library, which hides the complexity of the TCP/IP interface, offering a set of functions for establishing the connection (Connect and Disconnect), for synchronous transfer (Request, Poke), and for asynchronous transfer (AdviseStart and AdviseStop). The client must be able to intercept the RTDB events in an appropriate thread.

Recommended Reading

1. Wagner, F., "VFSM Executable Specification," *Proceedings of the International Conference on Computer System and Software Engineering*, The Hague, the Netherlands, 1992: 226–231.
2. Wagner, F., *The Virtual Finite State Machine: Executable Control Flow Specification*. Giessen: Rosa Fischer-Löw Verlag, 1994.
3. Wagner, F., Wolstenholme, P., "A modern real-time software design tool: applying lessons from Leo," *IEE Computing & Control Engineering* (February 2003).
4. Wagner, F., Wolstenholme, P., "Modeling and building reliable and re-usable software," *Proceedings of the ECBS'03*, Hunstville, April 2003.
5. Wagner, F., Wolstenholme, P., Wagner, T., "Closing the gap between software modeling and code," *Proceedings of the ECBS'04*, Brno, May 2004.

Chapter 11

Digital Input and Output

A Digital Input Has Three Control Values

We begin the discussion of RTDB objects with digital input (DI) and output (DO). Both objects are similar considering their values. In Boolean algebra they are treated as two values (false, true) signals. In the Vfsm concept, DI have three control values: LOW, HIGH, and UNKNOWN and there are two actions defined for DO: Low and High.

Figure 11.1 shows the state transition diagram of a DI object controlled by the external I/O signals Di_LOW, Di_HIGH, Di_UNKNOWN. We chose LOW and HIGH deliberately, although they are clearly not appropriate in all situations. They correspond to an electronic engineer's view of such signals, of course. One could argue for the more general names 0 and 1, or TRUE and FALSE, for example, but we wish to highlight the distinction between these values and binary Boolean values.

For the DI object the UNKNOWN control value represents missing information about DI. This means that effectively it plays a role mostly after a start-up; it may occur also if the hardware interface is down. In the latter case the I/O Handler must detect the hardware malfunction and set the UNKNOWN value in the RTDB using the Di_UNKNOWN signal. It means also that if we define only one control value for the DI object using either the LOW or HIGH value, the missing entry in VI is ambiguous — it may mean the not used or the unknown control value and we cannot use the absence of a control value for formulating conditions (a rule of our positive-logic algebra).

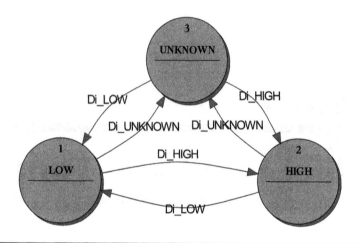

Figure 11.1 The DI Vfsm ST diagram.

Example

The example represents a pure combinational system with a digital input Di and a digital output Do. We use the table *Always* to define the Input Action expressions: input conditions and corresponding output actions.

Two Input Action expressions defined in Figure 11.2 say that LED (Do) is switched On (name used, e.g., for the value High) if Button (Di) is LOW and is switched Off (name used, e.g., for the value Low) if Button is HIGH.

Appendix O *DI_DO project* describes the details of this example.

| Always | Button_LOW | LED_On |
| Always | Button_HIGH | LED_Off |

Figure 11.2 Test_DI-DO example: the table *Always*.

Setting and Clearing the Boolean Output Are Two Different Actions

In the Vfsm concept each action has its governing command. To control Do we need two commands; the absence of one command does not guarantee that the other command is due. This feature has important

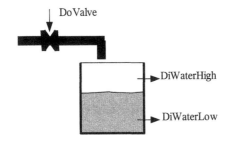

Figure 11.3   A water tank control.

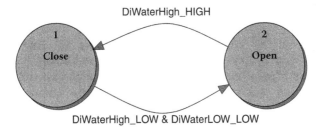

Figure 11.4   A state machine for the water tank control.

consequences, demonstrated with the example Tank, which is a valve control for a water tank (Figure 11.3).

The tank serves as a water supply buffer. The water level in the tank is monitored by two sensors. The water is supplied to the tank by a pipe equipped with an open/close valve. The valve is controlled by a signal DoValve; e.g., its High level opens the valve, the Low level closes it. If the tank is empty (DiWaterLow is LOW) the valve should be opened and it should stay opened until the tank is full (DiWaterHigh and DiWaterLow are HIGH). If the tank is full, the valve is closed and it should stay closed until the tank becomes almost empty (DiWaterLow is LOW again).

This simple valve control is a state machine (Figure 11.4) as for the input condition

`DiWaterHigh=LOW and DiWaterLow=HIGH`

the output can be High or Low depending on the previous input condition. Using a conventional automata theory design method we get an automaton with two states that requires one R-S flip-flop to encode the states. The flip-flop Set and Reset signals for a hardware design would be:

$$S = \overline{DiWaterHigh} \,\&\, DiWaterLow$$

$$R = DiWaterHigh$$

which in Vfsm convention reads

```
S = DiWaterHigh_LOW & DiWaterLow_LOW
R = DiWaterHigh_HIGH
```

In StateWORKS the state machine is superfluous as the DO object is already defined as such a state machine (see Figure 11.5).

Therefore, we may solve the water tank control by setting and resetting DoValve using actions: DoValve_Close and DoValve_Open as shown in Figure 11.6.

Analyzing a control problem most of us think in categories: the motor, pump, valve, etc. should be open if "here comes the open action" and should be closed if "here comes the close action," which exactly corresponds to the Vfsm concept. Of course, the open/close actions are not always so simple as in the example and to solve the control problem we do need a state machine. Anyway, the digital output understood and implemented as a flip-flop very often simplifies the solution.

In summary, in a situation where a single state machine is a solution, StateWORKS using the Vfsm concept presents a combinational solution for the control problem due to the flip-flop function of the DO object.

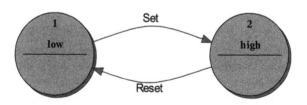

Figure 11.5   The DO Vfsm ST diagram.

Always	DiWaterHigh_HIGH	DoValve_Close
Always	DiWaterHigh_LOW & DiWaterLow_LOW	DoValve_Open

Figure 11.6   Tank example: the table *Always*.

DI and DO Properties

Both object types have only one Property: *Invert*, which defines whether the value is negated or not:

■ **Invert**: may be `false` (value = original) or `true` (value = negated).

A detailed description of object use and their properties can be found in Reference 1 through 3.

Recommended Reading

1. SW Software: StateWORKS. Reference Manual for the Class Library.
2. SW Software: StateWORKS Development Tools. User's Guide.
3. SW Software: StateWORKS Studio. Help.

Chapter 12

Other Inputs

Input Data (DAT)

A digital input is well defined and its control values correspond directly to its physical value. It is in this respect an exception; other input values with which we have to deal are not so homogeneous. For example, any data type can transport control information: integer, float, string, etc. To handle those data types the RTDB defines a basic data type **DAT,** which can be of any type typically used in programming languages. The object of DAT type is a container for any data, and it can be used, e.g., for storing strings.

Control Values

There are two types of control values generated by a DAT object: the changes of data and the data value. Although all numerical input objects possess these two sources of control value, for some of them the changes of data are important, and for others the data value is used.

The behavior of a DAT object is described by a state machine as in Figure 12.1. After start-up a DAT object is in the state **UNDEF**. The first data change forces it to go to the state **CHANGED**. If the DAT state is used by another object (e.g., as Input of a SWIP object — see the next section), it goes to the state **DEF**. From this time it loops between the states DEF and CHANGED: any data change forces it to go to the state CHANGED and if this event is used (consumed) by some other object it returns to the state DEF. DAT will return to the state UNDEF if data cannot be kept up to date. This may occur if a DAT object is used to store inputs delivered by an I/O

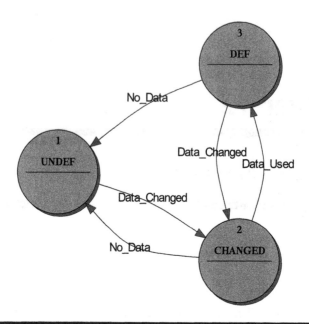

Figure 12.1   The DAT Vfsm ST diagram.

handler; if the I/O handler fails, it should set the DAT object to the UNDEF state using the signal No_Data. We may use the DAT states as control values.

In general, all input data objects (DAT, NI, PAR, UDC) have the states DEF and CHANGED; the rules of changing between them are those described for the DAT type. In practice there is a need for more specialized DAT objects. They may differ in initial states, which may be INIT or UNDEF and the role that the initial state plays. For some objects (PAR) the initial state is a true initial state, which once left is never reached again. Other objects (DAT, NI) may return to the initial state to signal a problem with a data update.

The other source of control value is the data value; i.e., the data stored by a DAT object may be of interest and we need a way to filter the control value from the object data. As there are several object types whose data contains control values, the RTDB has special objects SWIP (for numerical data) and STR (for string type) to do the job (the alternative would be to add this kind of feature to each object). If the SWIP or STR object supervises the data stored in a DAT object the DAT's own control value loses its sense and must not be used.

Example

The example shows the use of a data change as a source of a control value. In the example the state CHANGED of the object DAT is used to

Always	Dat_CHANGED	LED_Dat_Changed
Always	Di_HIGH	LED_Off

Figure 12.2   DAT example: the table *Always*.

switch on LED (LED_Dat_Changed means LED on). The digital input Di is used to switch off LED so that we can repeat the exercise (LED_Off means LED off). The control is specified in the table *Always* (Figure 12.2).

Appendix P: *Other_Inputs Project* describes the details of the project Test_DAT as an example.

Properties

The object of DAT type is a characterized by two properties:

- **Format**: any C-like data format (bool, short, unsigned short, etc.)
- **Unit**: any string (for instance V, A, mBar, m, sec, etc.)

The Unit property has an auxiliary use: it contains any string that defines a unit of a physical quantity and may be used for a display.

A detailed description of object use and their properties can be found in References 1 through 3.

Getting the Control Value (SWIP)

The **SWIP** object is a universal object used to supervise a data value in other objects. Data in objects like DAT may have an infinite number of values (theoretically). For control purposes only some of them are relevant described by ranges as in Figure 12.3.

Actions

To function, the SWIP object is controlled by actions (commands*), which are, on the one hand, changes of the supervised data (see Figure 12.3):

---

* We sometimes use the term *command* or *action command* as a substitute for *action* — it should not be confused with the function of the CMD object type.

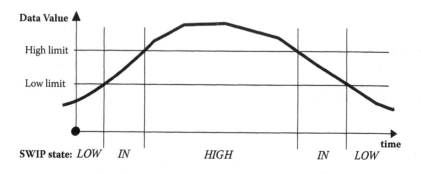

Figure 12.3 SWIP state reflects the control value of the supervised object.

- **in**: data value is in the range (Low limit >= Data value <= High limit).
- **high**: data value is larger than the High limit (Data value > High limit).
- **low**: data value is less than the Low limit (Data value < Low limit).
- **undef**: data value is not valid.

Note that a more complex supervision could be arranged by use of two SWIP objects to supervise the same data object, so as to have four boundaries rather than two.

On the other hand, SWIP is triggered by action commands received from a VFSM object:

- **Off**: command to disable SWIP
- **On**: command to enable SWIP

Control Values

The data value changes continuously in time. The SWIP state represents the control value of the supervised object according to the Limit High and Limit Low values (see SWIP properties later in the Properties section below). Figure 12.4 shows the state transition diagram of the SWIP object.

To start the supervision, SWIP must be switched on; i.e., the object leaves the state **OFF** and goes to one of the states: **HIGH**, **IN,** or **LOW**. The supervision loses its sense if the Input data is not valid — it occurs if the Input object switches to the UNDEF state. This situation is intercepted by the SWIP object, which then goes into the state **UNDEF**. All SWIP states may be used as control values. If the SWIP object supervises data stored in another object the control value of that supervised object loses its sense and must not be used.

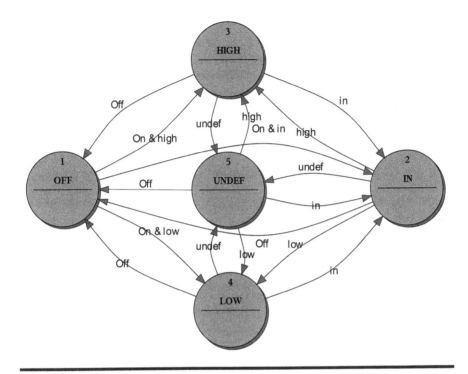

Figure 12.4   The SWIP Vfsm ST diagram.

Example

The state machine Test_SWIP (Figure 12.5) has a state *MyInit* in addition to the start-up state *Init*. The only purpose of this arrangement is to start the SWIP object in the Entry Action of the state *MyInit*. The entire control is defined in the table *Always* — see Figure 12.6. In the example, to signal the SWIP changes we use the output LED, which is on (LED_In) if SWIP is IN; otherwise it is off (LED_High_or_Low).

Appendix P *Other_Inputs project* describes the details of this example.

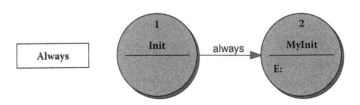

Figure 12.5   SWIP example: the ST diagram.

| Always | Swip_IN | LED_In |
| Always | Swip_HIGH | Swip_LOW | LED_High_or_Low |

Figure 12.6 SWIP example: the table *Always*.

Properties

The object of SWIP type is characterized by the following properties:

- **Input**: the name of the supervised object; the object type may be DAT, PAR, NI, or UDC.
- **Limit Low**: a number defining the lower value of the supervised range; may be a constant value or a parameter name.
- **Limit High**: a number defining the upper value of the supervised range; may be a constant value or a parameter name.

A detailed description of object use and their properties can be found in References 1 through 3.

NI Object as an Extension of DAT Type

The **NI** type is intended to store data coming from the outside world (hardware peripheral devices). It is like the DI object but for non-Boolean values.

Control Values

The behavior of an NI object can be described by the state machine of the DAT object as seen in Figure 12.1.

The use of an NI state as a control value is rather limited; the change of an NI value normally does not present any relevant control information. Much more interesting from the control point of view is the data value, which is transformed into the control value by means of a SWIP as described above for the DAT object. Supervising the range of the NI data by SWIP and the use of SWIP state for control is a typical application of NI objects (see Figure 12.7).

Properties

Containers that store data may require more properties than DAT so as to make the use of RTDB objects more convenient. In addition to **Format** and **Unit,** NI contains the following properties:

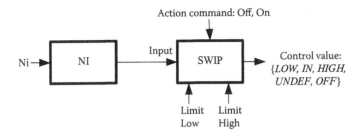

Figure 12.7   SWIP supervising NI.

- **Scale Mode**: none, Lin and Exp
  The *Lin* (linear) mode means that the original input data is scaled using the formula:

  $$y = a * x + b$$

  The *Exp* (exponential) mode means that the input original data is scaled using the formula:

  $$y = e^{a*x+b}$$

  where y = output, x = input, a = Scale Factor, b = Offset.
- **Scale Factor**: float or integer number
- **Offset**: float or integer number
- **Threshold**: any float or integer number

The requirement for those properties comes from hardware practice, which shows that analog input values represent various physical values like temperature, pressure, distance, speed, etc. These values are produced by sensors with certain characteristics (linear, exponential) and offsets. The analog values are converted by analog-to-digital (A/D) converters into numerical values, such as integer numbers. The outputs of A/D converters may have also offset or threshold. Of course, the user would like to see and use the actual value of a physical entity instead of a pure number. For instance, a value of 572 from an A/D converter is more difficult to comprehend than a value 36.5 mBar after scaling.

A detailed description of object use and their properties can be found in References 1 through 3.

PAR Object as a Specific Variant of DAT Type

The **PAR** type is intended to store data like the DAT object. This time data does not come from hardware, but represents values stored internally in software and used for defining some application-specific features; the data may be specified with a default Initial value and is called a parameter.

Control Values

The behavior of a PAR object is specified by a state transition diagram shown in Figure 12.8. The diagram possesses two initial states: **UNDEF** and **INIT**, which are used to differentiate between process parameters (PP) and equipment parameters (EP) (see PAR properties). The **DEF** and **CHANGED** state corresponds to similar states of the object DAT.

The PP parameter is valid if loaded by a start-up from a parameter file. If the parameter file is not loaded, the parameter has the default Initial value and the PAR object stays in the state UNDEF. The control may react to a problem during start-up (missing parameter file?) as a PP PAR object is initialized by a start-up to the state UNDEF.

The EP parameter is always valid as we assume that the system (computer) resources are available and if the EP parameter is not found in resources the Initial value is the intended choice. Thus, the EP PAR object is initialized to the state INIT by a start-up.

If the initial state (INIT or UNDEF) is left the PAR object never returns there.

The use of a PAR data value to generate control values is rather limited; if required it is done by SWIP, which gets the PAR object as Input. More interesting from the control point of view is the PAR state. Especially the state CHANGED is often used as a trigger to do some actions (see

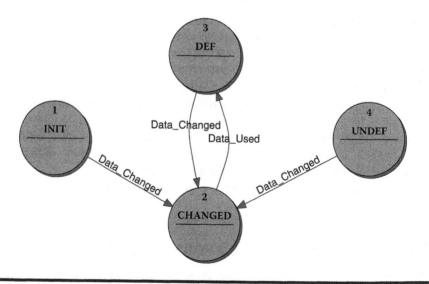

Figure 12.8 The PAR Vfsm ST diagram.

Example — Pressure supervision in Chapter 8) where the parameter change triggers the Output function). Reading of the PAR state "consumes" its control value and the object returns to the state DEF. Also in that case we should not forget that the PAR control values and the SWIP or STR control value cannot be used at the same time: if the SWIP or STR object supervises the data stored in a PAR object, the PAR control value loses its sense and must not be used (it will be "consumed" by SWIP or STR).

Properties

Also in this case we would like to have some additional properties to make the use of parameters as convenient as possible. In addition to **Format** and **Unit** as for the DAT object the PAR type contains the following properties:

- **Category**: PP or EP, several variants of each
- **Low limit**: any number
- **High limit**: any number
- **Initial value**: any number representing a default data value

The *Low* and *High limit* are numbers that may be used by the user, e.g., in a user interface; they do not play any role in the RTDB, and they are especially not used as limits for the Initial value. The *Initial value* is the default data value.

The most interesting property is the *Category,* which may belong to the PP or EP group.

The PP Category is intended for defining parameters for the actual process (in terms of controlled processes). In other words PP parameters are a kind of recipe, which may be changed from time to time. The way the PP parameters are changed is the user's responsibility. There are two variants of PP parameters: PP and PP_Coded where the latter is used to indicate that its value is generated in program code of an I/O handler or User Function.

The EP Category is used for defining environmental parameters — constant for a given application in a given system. In principle, EP parameters do not change when the application runs. The EP parameters are stored in system resources (Registry under Windows or a special file) and loaded during start-up from the resources. If there are no entries in the resources, the Initial value is taken. There are three variants of EP parameters: EP, EP_LM_USERS, and EP_LM_ADMIN whose utilization depends on user rights (single user, all users, and administrator).

A detailed description of object use and their properties can be found in Reference 1 through 3.

String (STR) as a Specific Variant of SWIP

The DAT object may contain data of string type. As SWIP cannot deal with such a non-numerical problem, we have the **STR** object as a supervisor of string data. For identifying the string content, the STR object compares the Input (see STR properties) string with one or more predefined patterns. The pattern(s) to be matched must be specified as a regular expression (RE) as known and used in the software, especially UNIX, world. In addition to the result control value (MATCH, NOMATCH, ERROR) the Input string is partitioned into substrings according to the RE and placed in specified data objects; in case of NOMATCH, the data values of the destination objects do not change. The substrings are converted to the destination types.

Actions

Like similar objects (see, e.g., SWIP) STR is controlled by action commands. There are explicit commands available as actions:

- **Off**: disables the object.
- **On**: enables the object.
- **Set**: activates the matching operation.

and internal commands generated by the matching process:

- **match**: the Input string matches RE.
- **nomatch**: the Input string does not match RE.
- **error**: RE is erroneous.

Matching starts if STR is enabled and either the Input string or the RE changes. When matching is finished, STR is deactivated and can be activated by the command *Set*.

Control Values

The behavior of a STR object is specified by a state transition diagram shown in Figure 12.9.

On start-up STR is in the state **OFF**. The command *On* enables its operation: the object goes to the state **INIT**. After the first match operation STR goes to one of the states: **MATCH**, **NOMATCH**, or **ERROR**. The command *Set* activates the object — STR goes to the state **DEF**, where it is able to perform the next matching operation. From that time it oscillates between

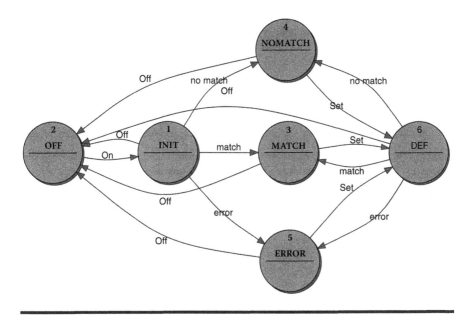

Figure 12.9   The STR Vfsm ST diagram.

one of the MATCH, NOMATCH, ERROR states and the DEF state comparing the Input string with RE and being activated by the Set command for the next matching operation. Note that the matching operation is triggered by a new Input string; the same string written into Input is ignored.

As for SWIP, if the STR object supervises data stored in another object, that object's control value loses its sense and must not be used.

Example

The example demonstrates a simple application where the STR object is used to detect a string "Start" or "Stop" plus a number of the device to be started/stopped. The state machine has two states: *Init* and *MyInit*; the second one is for enabling the STR object. The entire control we want to show is the Input Action in the state *MyInit* (see Figure 12.10).

On entering the state *MyInit* the state machine sends a command *On* to the STR object, which enables its activity — the object is in state INIT. Any change of the Input string triggers the matching process and the STR goes into one of the states: NOMATCH, MATCH, or ERROR depending on the matching result. Receiving in that moment the command Cmd_Set the state machine sends the command Set to the STR object, which goes into the state DEF. In that state it reacts to changes of the Input string similarly as in the state INIT.

MyInit	Entry action	Str_On
	eXit action	
	Cmd_Set	Str_Set MyCmd_Clear

Figure 12.10   STR example: the ST table of the state *MyInit*.

To "consume" the command Cmd_Set the state machine uses the signal MyCmd_Clear, which clears the present command allowing a repetition of the command.

Note that to fully understand the functioning we have to look at the properties. For instance, the essential role played here by the regular expression, which defines the matching string:

RE = (Start|Stop[ ]+)([0-9]+)

Appendix P *Other_Inputs project* describes the details of this example.

Properties

The STR object has the following properties:

- **Input**: the supervised object; may be DAT or PAR.
- **Regular expression**: defines the RE; may be a string or the name of a DAT or PAR object, which contains the RE.
- **List of substrings**: a list of destination objects (which may be DAT, PAR, NI, or NO), which receive the substrings produced by matching. The strings are automatically converted into the format of the destination objects.

STR objects can be quite complex. The ability to analyze strings using regular expressions and to assign the results to a number of single-valued objects makes STR a very powerful device for message-based applications.

A detailed description of object use and their properties can be found in References 1 through 3.

Recommended Reading

1. SW Software: StateWORKS. Reference Manual for the Class Library.
2. SW Software: StateWORKS Development Tools. User's Guide.
3. SW Software: StateWORKS Studio. Help.

Chapter 13

Other Outputs

Output Data (NO)

Setting a digital output is the simplest and most often used operation, especially in industrial control. In general, we would like to output any value. For this purpose the RTDB has an object type **NO** for numerical output. The NO object is a kind of switch (see Figure 13.1 and the Actions as described below) used to pass the *Out Data* defined in a DAT or PAR object to the output. The TAB object may be used to expand the possible sources of output data. We may also use NI and UDC objects as the source of NO *Out Data,* but that would be very unusual. One should note that a value could have been set into a DAT or PAR object in the RTDB by some other program, external to the StateWORKS system, if that were required for the application.

Actions

The behavior of NO may be described by a state machine diagram as in Figure 13.2, where the transitions are triggered by action commands:

- **Off**: sets the value 0 to the output.
- **On**: links the Out Data object (PAR, DAT, TAB) continuously with the output.
- **Set**: sets the Out Data object (PAR, DAT, TAB) to the output; the value is valid until changed by another *Set* command.

221

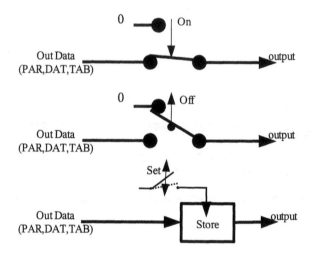

Figure 13.1   NO as an output switch.

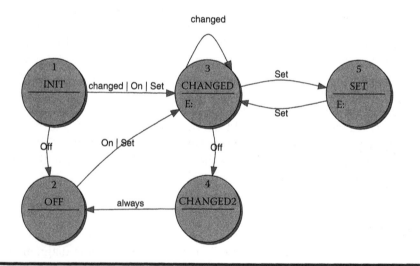

Figure 13.2   The NO Vfsm ST diagram.

The signal generated by the NO object itself is:

■ **Changed**: signals that Out Data has changed.

The state transition diagram explains the way this object type can be used. The state transition diagram does not play any role in generating control values as it would be rather strange to use the NO state for control purposes.

After start-up, the NO object is in the state **OFF** if its *Out Data* is a PAR or DAT object — in this state NO is disabled passing the value 0 to its output. If NO *Out Data* is the TAB object the NO object is after start-up in the state **INIT** — in this state NO is enabled and any data change forces it to go to the state **CHANGED**. If we leave it there, it works as if it has received the On command passing any *Out Data* (PAR, DAT, TAB) to its output. It is the most common way of using an NO object and the object does not require any initialization. Applying Set command stores the current value of *Out Data* (PAR, DAT, TAB) in the NO object, and from that moment this data is passed to the output until a new Set command actualizes it.

With the Off command we force the NO object to go to the state OFF where it cuts off the output from the *Out Data* and sets the value 0 to the output. The transition from the state CHANGED to OFF is done via a CHANGED2 state, which is transparent to the outside — it is seen as CHANGED. Receiving the On command the NO object returns to the CHANGED state.

Receiving the command Set in any state, it goes to the state **SET** via the state CHANGED. This applies also to the state SET: receiving the command Set, it goes to the state CHANGED and returns immediately to SET. An *Out Data* change while NO is in the state CHANGED causes a transition to the same state — CHANGED. Therefore, to avoid a loop the signal *changed* is cleared in the state CHANGE in the Entry Action. Similarly, the Set command is cleared in Entry Action of the state SET.

Example

Thus, there are several imaginable schemes to set the NO output. In practice, we choose one of two possible variants:

- Enabling the NO object by sending the command On — from this moment the value on its *Out Data* is passed to the output.
- Sending the NO object *Out Data* to the output by sending the command Set — actualization of the output requires a repetition of the command.

In Appendix Q *Other_Outputs project* there is an example that demonstrates the possible use of the NO object.

Properties

The NO object has properties similar to NI allowing convenient cooperation with digital-to-analog (D/A) converters, which are the primary destinations of NO data. The properties of NO are as follows:

- **Format**: any C-like data format (bool, short, unsigned short, etc., except string)
- **Unit**: any string (e.g., V, A, mBar, m, sec, etc.)
- **Scale Mode**: none, Lin, and Exp

  The Lin (linear) mode means that before passing it to the output the original *Out Data* is scaled using the formula:

  $$y = a^*x + b$$

  The Exp (exponential) mode means that before passing it to the output the original *Out Data* is scaled using the formula:

  $$y = e^{a^*x+b}$$

  where $y \rightarrow$ output, $x \rightarrow$ Out Data, $a \rightarrow$ Scaling Factor, $b \rightarrow$ Offset.
- **Scale Factor**: float or integer number
- **Offset**: float or integer number
- **Out Data**: the data to be output; may be DAT, PAR, NI, UDC, or TAB objects

A detailed description of object use and their properties can be found in References 1 through 3.

Output Demultiplexer (TAB)

The **TAB** object selects one object from several to derive its output. It is used as an output in the I/O handler (see Figure 13.3). As TAB input objects we can use PAR, DAT, and NI types. To realize it TAB has a property:

- **Table rows**: specifies the used input objects.

TAB is used primarily for delivering *Out Data* to an NO object if they are to come from various sources as in the project Other_Outputs discussed in the previous section. TAB can be used also as an independent output. A good example of such use is passing several strings to the I/O handler.

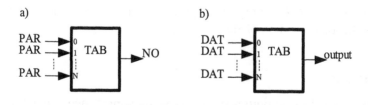

Figure 13.3   TAB object (a) delivering PAR values to NO; (b) sending values (strings) to I/O output directly.

Actions

The actions controlling TAB switching are just numbers which are indices 0 ... N to its table of input objects.

A detailed description of object use and their properties can be found in References 1 through 3.

Example

The project Other_Outputs also contains a Tab object of type TAB. Appendix Q *Other_Outputs project* describes the details of this example.

Alarms (AL)

An **AL** object stores messages called Alarms. An Alarm carries no control information — it is only a message about a usually erroneous situation, but in general, it may contain any information. Each single Alarm, as well as all Alarms in the system, requires quite sophisticated handling to make Alarm use convenient for the users.

Actions

The behavior of the AL object is well described by a state transition diagram in Figure 13.4. AL is controlled by action commands:

- **Staying**: forces a generation of a Staying Alarm.
- **Coming**: forces a generation of a Coming Alarm in expectation of a following Going Alarm.
- **Going**: forces a generation of a Going Alarm, which is to signal that the Alarm reason has disappeared.

and an external signal:

- **Ack**: passed directly to the AL object (cannot be specified as an action).

While specifying an Alarm as an output its action command may then receive one of three commands: Staying, Coming, or Going. We use the command Staying if we decide that the reason for the Alarm is irreversible and will not disappear by itself. The other two commands, Coming and Going, must be used in pairs. We use the Coming command if we assume that some time later we will use the Going command to signal that the reason for the Alarm has vanished.

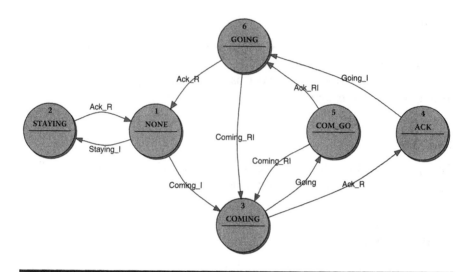

Figure 13.4   The AL Vfsm ST diagram.

In addition to the Alarm objects the RTDB has a pseudo-Alarm (with the name AL equal to the type name), which is a list of all not-yet-acknowledged alarms. Any Alarm is entered at the top of the list and is accessible (visible). In the state transition diagram the I and R characters accompanying the transition conditions denote the Transition Actions performed on a state change: I stays for "Insert the Alarm into the AL queue" and R stays for "Remove the Alarm from the AL queue."

The pseudo-Alarm AL is "intelligent" enough to reject a new Alarm if it is already in the list. This makes the use of Alarms more comfortable.

Properties

The AL object has two properties:

- **Category**: positive integer number. Defines the severity: Error, Warning, and Information are the typical values. A user may define any specific value. The meaning of the Category value is shown in Table 13.1.
- **Text**: alarm string to be displayed.

The Category expresses only the intention; the actually logged Alarms depend on the *AL_CatKeyPar* EP parameter, and those Alarms are logged in the AlarmLog.txt file. The parameter *AL_CatKeyPar* may have the values 0 ... 7 with the following binary coded meaning:

- 1 – write `Errors` to AlarmLog
- 2 – write `Warnings` to AlarmLog
- 4 – write `Informations` to AlarmLog

For instance, the value 1 means that all Errors will be logged and value 7 (1 + 2 + 4) means that all three Alarm categories: `Error`, `Warning`, and `Information` will be logged. If the *AL_CatKeyPar* EP parameter is not defined, all Alarms are logged. Alarms with user-defined categories other than 1, 2, or 4 are not logged.

Table 13.1 Alarm Severity Logged in AlarmLog.txt

Category	Coming, Staying	Going, None
1	Error	Information
2	Warning	Information
4	Information	Information
Other	User defined	User defined

The Text property is used to define the text to display if the Alarm is generated. The Text may contain:

- Pure text
- Identifier beginning with `IDS_`
- Name of an RTDB object (DAT, NI, PAR, NO, UDC) beginning with %

Any combination of the three string forms is allowed by building the Text string; space is used as a separator. The `IDS_` identifiers can be prepared and are managed by StateWORKS Studio.

The Text string is then completed by Alarm Output value defined in the state machine, which uses the object and time taken from computer resources. Such an Alarm text can be fetched as a whole (attribute `AIL`) or as separate attributes one by one (Type, Text, Category, Time).

A detailed description of object use and their properties can be found in References 1 through 3.

Example

Exercises with the state machine Test_AL in the Appendix Q *Other_Outputs project* demonstrate the Alarm possibilities.

For the three alarms used, the Text and Category have been defined as follows:

- Al:01 *Text*: This is Al:01 alarm   *Category*: 1
- Al:02 *Text*: No:01 has changed:. % No:01   *Category*: 2
- Al:03 *Text*: IDS_TEXT   *Category*: 4

Text 13.1   Example of the AlarmLog.txt file

```
C:...\Other_Outputs\Conf\Other_outputs.swd started
at: 05-Nov-04 12:56:19
ERROR    05-Nov-04 12:56:33 - A1:01 - COMING - This is A1:01 alarm
INFO     05-Nov-04 12:56:34 - A1:01 - COM_GO
WARNING  05-Nov-04 12:56:37 - A1:02 - STAYING - No:01 has changed: 20mBar
INFO     05-Nov-04 12:56:53 - A1:03 - STAYING - This text has been taken from Resources
INFO     05-Nov-04 12:56:55 - A1:03 - NONE
INFO     05-Nov-04 12:56:56 - A1:02 - NONE
INFO     05-Nov-04 12:56:57 - A1:01 - GOING
INFO     05-Nov-04 12:56:57 - A1:01 - NONE
```

After some experiments: generating Alarms and acknowledging them the AlarmLog.txt file contains entries as shown in Text 13.1. We see that the %No:01 name has been replaced by the actual data value of the object No:01 equals 20mBar and the IDS_TEXT identifier has been replaced by the string "This text has been taken from Resources."

The possibility of using a text identifier allows internationalization of the Alarm text. By supplying stringres.src files with text in different languages we can adapt the displayed information to the language used.

Recommended Reading

1. SW Software: StateWORKS. Reference Manual for the Class Library.
2. SW Software: StateWORKS Development Tools. User's Guide.
3. SW Software: StateWORKS Studio. Help.

Chapter 14

Counters

A Simple Counter (CNT)

Counters are very often used in the design of state machines. A counter may be forced to increment or decrement its counting register by commands (actions) or it may count some events. Therefore the RTDB has several counter types, which simplify certain control problems. Counting itself is normally of no relevance in guiding the control. The only interesting moment is when a certain counting value, called an *expiration value,* is reached.

The simplest counter is represented by an object **CNT**.

Actions

The CNT object is controlled by action commands:

- **Reset**: resets the counting register
- **Start**: enables counting
- **ResetStart**: resets the counting register and enables counting
- **Stop**: disables counting
- **Inc**: increments the counting register
- **Dec**: decrements the counting register

and a CNT internal command:

- **Expired**: generated if the counting register equals or is larger than the expiration value

An object of a type CNT counts after start-up from 0. Typically, a CNT object is enabled by the ResetStart command and counts forward on receiving the command Inc, and when its counting register reaches the expiration value, it generates the overflow signal.

Control Values

The behavior of a CNT object is specified by a state transition diagram shown in Figure 14.1. A CNT object waits in the **RESET** state to be enabled. Receiving the command Start or ResetStart it goes to the state **RUN** where it is able to react to the command Inc, which increments its counting register, and to the command Dec, which decrements its counting register. While in the state RUN a CNT object may be disabled again by a command Reset or Stop. In the state **STOP** a CNT object is disabled but it may resume counting from the last value (receiving command Start) or counting from 0 (receiving the command ResetStart). The command ResetStart in the state RUN restarts counting from 0.

Any state may be used as a control value but the most interesting state is the state **OVER**, which signals that the counter has reached the expiration value. While in the state OVER the counter continues counting of commands Inc and Dec. The command Stop may disable counting in OVER forcing the transition to the state **OVERSTOP**. From both OVER states CNT may be reset or start counting from 0.

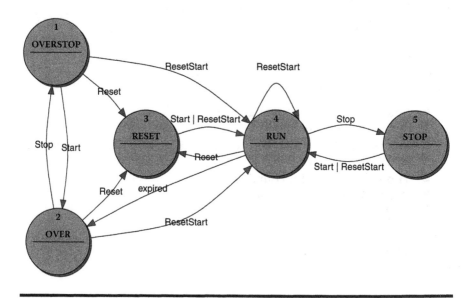

Figure 14.1   The CNT Vfsm ST diagram.

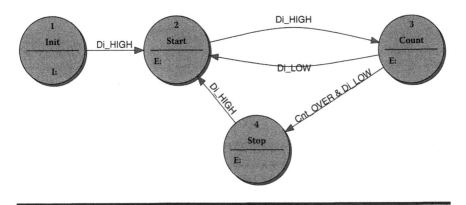

Figure 14.2   The state machine counting Di changes.

Example

The CNT object has to count changes (to HIGH) of a digital input Di. We use for this purpose the state machine Test_CNT shown in Figure 14.2. After start-up the state machine is in the state *Init*. The first change of Di to HIGH starts the counter (in the Input Action) and changes the state to *Start*. On entering the state *Start* the output LED is switched off (at that first entry it has been initialized to off). The state machine continues the state changes (Di is HIGH) going to the state *Count* where it sends the Inc command to the counter. On Di LOW it returns to the state *Start*. The next Di HIGH means a transition to the state *Count* and counter increment and so on. This switching between these two states lasts until the counter reaches the expiration value and goes to its state OVER. The counter OVER causes a transition of the state machine to the state *Stop* where LED is switched on signaling the expiration and the counter is restarted. The next Di HIGH causes a transition to the state *Start* and the counting will be repeated.

We have to be careful in specifying the transition in the state *Count* (see Figure 14.3): the pure transition condition Cnt_OVER to the state *Stop* would not work. It will cause an immediate transition to *Count* if Di is HIGH (which generates Cnt_OVER). This in turn will cause a further transition to the state *Start* and the state machine Test_CNT will loop through the states: *Start–Count–Stop*. If we use the Cnt_OVER and Di_LOW condition, the transition to the state *Stop* is delayed until Di is LOW. We note also that the sequence of Transition expressions in the state transition table is relevant.

Appendix R *Counters project* describes the details of this example.

We would not recommend that anyone solve this type of counting problem in this way — using the ECNT object offers a much simpler

Count	Entry action	Cnt_Inc
	eXit action	
Stop	Cnt_OVER & Di_LOW	
Start	Di_LOW	

Figure 14.3 CNT example: the ST table of the state *Count*.

solution. At least, this example well illustrates the use of the CNT object by requiring:

- Enabling and resetting the counter
- Incrementing the counter by command

Properties

The object of CNT type has three properties defining the expiration value:

- **By value**
- **Object name**
- **Const Value**

Expiration value equals *Const Value* if *By value* is true; otherwise Expiration value equals the value of an object (PAR) defined under *Object name*.

A detailed description of object use and properties can be found in References 1 through 3.

An Event Counter (ECNT)

The simple counter (CNT object) "counts" its commands Inc and Dec. Indirectly, it may count several things. We have shown in the CNT example how to count a state being triggered by a command Inc on entering the state — indirectly CNT counts the changes of a DI object. This solution was over complex — the ECNT object makes life much easier.

The **ECNT** object is very like the CNT object. Both object types have the same Actions and Control values; they differ only in Properties.

ECNT has exactly the same commands as the CNT object. The use of actions is similar but the Inc and Dec commands are rarely used as the ECNT task is to count control values of other objects and not its own commands.

The state transition diagram of the CNT object applies also to the ECNT object and the state OVER is the most interesting and commonly used control value.

Example

The state machine Test_ECNT realizes the same problem as the previous example for the CNT object: it counts the changes (to HIGH) of a digital input and the expiration is signaled by the output LED. In addition, we use a command Cmd_RestartEcnt to restart the counter.

In that case we have solved the problem in the table *Always* shown in Figure 14.4. The command Cmd_RestartEcnt switches off the LED and restarts the ECNT object. Expiration (Ecnt_OVER) of the counter switches on the LED and clears the command — the command Cmd_RestartEcnt may start counting again.

Always	Cmd_RestartEcnt	LED_Off Ecnt_Restart
Always	Ecnt_OVER	LED_On MyCmd_Clear

Figure 14.4 ECNT example: the table *Always*.

The object whose changes are counted is defined in the ECNT properties. This means that the state machine itself does not contain the complete control information. It controls the ECNT object (sending commands) and uses control values generated by the object but the true counting circumstances are hidden and can be seen in the project specification. In other words for the state machine Test_ECNT the object ECNT is a Slave state machine.

Appendix R *Counters project* describes the details of this example.

Properties

The object of ECNT type has properties defining the *Const value* (see CNT properties) and two additional properties:

- **Input**: defines the object whose control value (state) is to be counted. It may be an object of any type except CMD and UNIT.
- **Up Value**: defines the control value to be counted.

A detailed description of object use and properties can be found in References 1 through 3.

A Timer (TI)

A timer (**TI**) is a very frequently used object type. Especially, when designing state machines controlling directly peripheral devices we use timers to guard against deadlocks. An object of type TI is a specific event counter — it counts clock pulses, effectively measuring time elapsed since its start.

TI has the same action commands as the CNT counter except for Inc and Dec, which of course would not make sense.

The state transition diagram of the CNT object applies also to the TI object and the OVER state is the most interesting and used control value. The expiration value for TI is called a **time-out**.

A typical use of the TI object is shown in a state transition table in Figure 14.5. On entering the state *Busy* the state machine does something and waits for the reaction. At the same time it starts a timer so that it does not remain forever in that state. If the expected reaction comes before the timer expires, the state machine goes to the state *Done*. Otherwise, the time determined by the time-out value elapses and the state machine goes to the state *Error*. In most cases OVER only has a local value — after leaving a state it becomes insignificant. Therefore, as a rule the timer is stopped on exiting the state.

Busy	Entry action	DoSomething Ti_ResetStart
	eXit action	Ti_Stop
Done	ExpectedReaction	
Error	Ti_OVER	

Figure 14.5 A TI object guarding a busy state.

Properties

The object of TI type has properties defining the *Const value* (see CNT properties). As the counting object is by definition determined, it has only one additional property:

■ **Clock**: defining the clock period that may be: min, sec, 100msec.

The first "tick" of the clock is not precisely defined. Therefore, we recommend for small time-out values that a clock period one level lower than required be used, e.g., instead of 1 sec it is better to use 10*100msec.

A detailed description of object use and properties can be found in References 1 through 3.

An Up/Down Counter (UDC)

The ECNT object is powerful and its features are sufficient in most cases. The weak point of ECNT is in counting events only in one direction — a bidirectional counter may be required sometimes. For such applications the RTDB has another up/down event counter of type **UDC**. An object of UDC type is completely different than the up-to-now discussed family of counters. UDC is an example of an object with very powerful functionality considering its behavior as well as its properties.

Actions

An object of type UDC has the following action commands:

■ **Clear**: to clear its counting register
■ **Up**: to increment its counting register
■ **Down**: to decrement its counting register

Those commands correspond to Reset, Inc, and Dec of CNT and ECNT counters. The different names have been chosen to underline the different character of the UDC counter.

Control Values

UDC control values are the same as those of EP PAR (see state transition diagram in Figure 12.8) but these are not very often employed. An object of UDC type is rather exotic and its task is to perform some unusual counting function where its behavior is irrelevant.

UDC does not have an expiration value like other counters. Instead we have to use SWIP to recognize the counted value. This opens some different perspectives as shown in the example below. For example, several counter values may be detected and both positive and negative integers can be supervised. But first we have to present the properties, which are also completely different from those of other counters.

Properties

An object of UDC type has the following properties:

- **Unit**: any string for display purpose
- **Up Input**: the object name to be used for forward counting
- **Up Value**: the control value to be counted
- **Down Input**: the object name to be used for backward counting
- **Down Value**: the control value to be counted
- **Clear Input**: the object name to be used for clearing the counting register
- **Clear Value**: the control value used as a clear value

The UDC counts changes of control values of the declared objects. Any object type except CMD and UNIT may be used for this purpose. The control values may be used to count up, down, or clear the counting register (set to 0).

A detailed description of object use and properties can be found in References 1 through 3.

Example

The example Test_UDC demonstrates some possibilities of the UDC. The essence of the example lies in the property definitions. Analyzing them we see how it works:

- It counts forward (Up) changes of the state Stop (4) of the state machine Test_CNT.
- It counts backward (Down) the value HIGH (1) of the digital input Di:UdcDownInput.
- It can be cleared by the value LOW (0) of the digital input Di:Udc-ClearInput.

- A switchpoint Swip_for_UDC:01 detects the value 5 of the counting register.
- A second switchpoint Swip_for_UDC:02 detects the range between -2 and 3, with the limits defined by parameters Par:UDCLimitLow and Par:UDCLimitHigh, which could, of course, be changed to suit different circumstances, and even dynamically by some software process able to access the RTDB.

Appendix R *Counters project* contains the details of this example.

Recommended Reading

1. SW Software: StateWORKS. Reference Manual for the Class Library.
2. SW Software: StateWORKS Development Tools. User's Guide.
3. SW Software: StateWORKS Studio. Help.

Chapter 15

VFSM and Its Interfaces

Virtual Finite State Machine Interfaces

VFSM is the central object in the RTDB, storing the behavior specification and allowing the Vfsm Executor to carry out the specification. To realize that task the VFSM object has links to other objects. The diagram in Figure 15.1 shows all participating interface elements. There are a few interfaces between:

- State machines: realized by CMD objects and a direct access to VFSM states (the "state" links between VFSM are shown using thick lines)
- State machines and I/O handlers: realized by UNIT objects
- State machines and output functions (via OFUN objects): realized by UNIT objects

The diagram does not show objects owned by VFSMs (except the CMD). The VFSM object list contains by definition a command (CMD) object as it is assumed that a VFSM should have a command. In practice, when specifying simple examples we design also state machines without a command — this is accepted by the runtime system. State machines designed for a system of state machines have commands as the commands realize the interface among them.

We also mention in passing that a UNIT for OFUN may be replaced by VFSM, which owns objects required by the output function (both: UNIT and VFSM are lists of object names).

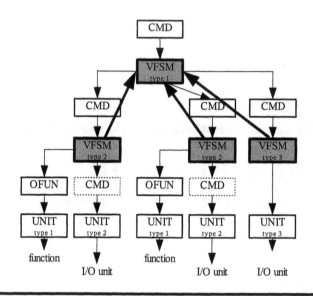

Figure 15.1   VFSM interfaces.

A Virtual Finite State Machine (VFSM)

A **VFSM** object is an instance of a state machine specified in a virtual environment. A VFSM type is just a list of objects used by a Vfsm specification — the list represents VFSM **properties**. VFSM behavior is defined by the specification, which depends on the just developed application. Hence, VFSM is a general term defining a group of VFSM types in a project. We should note the difference: there are several types of objects like CMD, DI, NI, SWIP, etc., but VFSM defines an entire group of very different objects, each described by its own state transition diagram. Each VFSM type may have of course several instances — objects of a given VFSM type like any other object types. For instance, the system in Figure 15.1 contains three different VFSM types: *type 1, type 2,* and *type 3,* with *type 2* used twice.

The interface between state machines is a command–state one; i.e., one state machine (a Master) sends a **command** to another state machine (a Slave) as shown in Figure 15.2. The Slave in turn displays to the Master its **state,** which is used by the Master as a control value. To realize the interface we need a CMD object, which is discussed in the following chapter. Here we talk only about the "state" interface, which passes the control information from Slave state machines to the Master.

Specifying the behavior of a Master state machine VFSM *type 1* we declare its Slaves: two VFSM *type 2* and one VFSM *type 3* as its I/O Inputs.

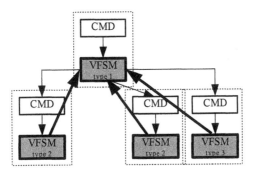

Figure 15.2   VFSM CMD–state interface.

From that moment the Slave states can be used as control values. When specifying (Transition or Input Action) conditions we are confronted very often with the following problem: something should be done for several Slave states. A direct specification of such requirements results in rather long expressions, which make the ST table more difficult to read. The Vfsm editor allows here a short cut — instead of writing, for example:

```
( Pressure1_Starting | Pressure1_Regulating ) &
( Pressure2_Starting | Pressure2_Regulating )
```

as we have done in the example *Pumps* (see Figure 9.12) we could define a name `Pressure_StartingORRegulating` combining that logical expression under one control value. The editor allows creation of **complex logical conditions** according to the formula:

```
Compact_Name =
(Slave1_State1 | Slave1_State2 | ... Slave1_StateN) &
(Slave2_State1 | Slave2_State2 | ... Slave2_StateN) &
...
(SlaveK_State1 | SlaveK_State2 | ... SlaveK_StateN)
```

This feature leads to very compact expressions, but we should not fail to invent expressive names that remind us what the control values mean. A compact specification is good if it explains the behavior well — hiding details of the specification is a very bad habit. For that reason the facility is limited to groups of state names, and is not applicable to all object types.

Hiding Specification Details

The use of control values in the form of complex expressions described above is not a completely straightforward issue. In some cases, it really makes the specification more readable, placing under a nice name a complex expression. But we should not forget that such a control value hides the details of the specification. Thus, we have to look at the full expression to understand completely the state machine behavior. Therefore, the idea needs very careful use. We should do it only if we really have the feeling that hiding details makes sense in that case.

It is not the only point where we may hide details. Using the event counter (ECNT) we do not see in the Vfsm specification what events are counted by the counter — they are not defined and seen until the RTDB object properties for each specific ECNT are specified. In certain situations an alternative solution using an object of a CNT (see *A simple counter (CNT)* in Chapter 14) type is clearer showing explicitly when the counter increases its value with the action Inc. The evaluation of what is better is in a sense a matter of taste. It is also a compromise between two factors, which influence the comprehensibility: compactness or explicitness.

Similarly, when supervising a value of NI or PAR with a SWIP object we do not see the supervised object in the Vfsm specification — it is determined in the system configuration process. That decision is made by specifying RTDB objects.

The possibility of hiding some specification details is on the one hand useful, making the specification more readable or at least we get a superficial feeling that it is more readable. On the other hand, we should like to have all behavior details in the specification; hiding details makes things eventually less comprehensible. Taking into account that hiding is unavoidable in some cases (e.g., using SWIP objects) we have to live with it but hiding details should not be a goal of our specification effort. Note that the approach does not conform to other software practices but it results from the basic definition of the Vfsm concept: to show the entire behavior as a function of control values.

A Command (CMD)

The specific role of **CMD** as a link between two state machines requires special management. When specifying state machines we are confronted with two "types" of CMD objects (see Figure 15.3). For some state machines (particularly for a Master on the hierarchy top level), commands will come from the outside world; e.g., they could be set by an operator or by some software process, even in another computer system over a network. Commands for most state machines are set by their Master state machine. In the second aspect the CMD object is controlled by actions of a Master state machine.

Because of this double access feature, the Vfsm editor "knows" about two CMD object types: **CMD-IN** and **CMD-OUT**.

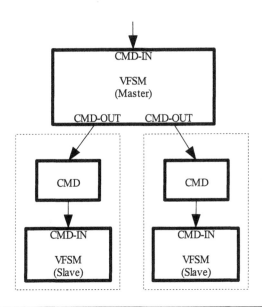

Figure 15.3 **A CMD object as the interface between a Master and Slaves state machines.**

The CMD-IN object is the state machine command as set by an operator or other state machine; normally each state machine has one CMD-IN and its assigned default name is **MyCmd**. The CMD-OUT is a foreign command declared in another state machine; normally a Master state machine has a few of them — one for each Slave. As most state machines are both a Slave and a Master, in general a state machine has its single CMD-IN object and several CMD-OUT objects defined by its Slaves.

Actually, there is only one object type CMD. The two notations used in the Vfsm editor are to avoid confusion and to make it easier to differentiate between the input commands (CMD-IN), which the state machine gets from its Master, and the output commands (CMD-OUT), which the state machine sends to its Slaves (if a state machine is both a Master and a Slave).

So, we have learned that any state machine has by definition a CMD-IN. We may say that a CMD-IN is owned by the state machine. A designer may introduce additional CMD-IN objects. A possible reason could be, e.g., to have a set of commands for maintenance that the designer wants to separate from Master access. In practice additional CMD-IN are seldom used. The auxiliary role of additional CMD-IN results also from the fact that a Master state machine "sees" and can send commands only to the default MyCmd of a Slave's object. Hence, an additional CMD-IN can be only used to send commands directly to a state machine from a Monitor or operator.

In behavior specification a CMD-IN object is, of course, a source of control values and we may define Input names on command names (values). A CMD-IN has also one output value `Clear`, which can be used to define an action. Here we discover the problem of the command lifetime (see also section *Signal lifetime* in Chapter 7). A command is a signal which a designer invents to organize the cooperation of state machines and it is the designer's responsibility to decide how long a concrete command value is valid. The command value may be needed for Transition or Input Action conditions in several consecutive states or is not used and just irrelevant but eventually it is replaced by another command. Sometimes, a state machine does a loop through several states returning to the state that has been left due to a command — if the command is still there, the state machine will continue the (in such a case infinite) loop. The value `Clear` is then a very useful command allowing a definition of an action, which sets the command to a value 0 (hence, the value 0 must not be used as a command and to be on the safe side the Vfsm editor prevents it). As the state machine owns the CMD-IN object, it is logical that this state machine has the chance to determine the lifetime of the command — it does it when it does not need the command any longer or when it must do it to avoid undesirable effects as in the infinite loop example.

Figure 15.4 shows the two possible ways of handling a command. Receiving the `Cmd_Start` the state machine Test_CMD goes to the state *Start*, performs there the required Entry actions, and returns to *Init* on receiving the command `Cmd_Stop` (Figure 15.5). Similar behavior can be

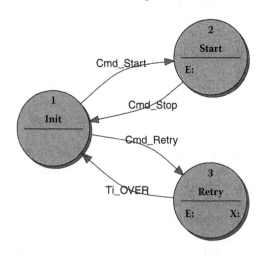

Figure 15.4 Test_CMD: the ST diagram.

Start	Entry action	DoSomething
	eXit action	
Init	Cmd_Stop	

Figure 15.5   Test_CMD: the ST table of the state *Start*.

The timer causes a break which is required to clear the Cmd.

Retry	Entry action	DoSomething MyCmd_Clear Ti_ResetStart
	eXit action	Ti_Stop
Init	Ti_OVER	

Figure 15.6   Test_CMD: the ST table of the state *Retry*.

achieved using the Cmd_Retry which switches the state machine into the state *Retry*. On entering the state *Retry* the state machine performs the required Entry action and returns to the state *Init* (Figure 15.6). To avoid an infinite loop the command value has been cleared on entering the state *Retry*. A timer is used to cause a required break to assure that the command is cleared.

A Master state machine defines actions using commands of its Slaves (seen as CMD-OUT). The comprehensively discussed project Pumps is a good illustration of the Master–Slave links.

A special problem is presented by a command link between state machines and I/O handlers. Although an I/O handler is not a Vfsm, it may be controlled by commands. In Figure 15.1 the VFSM *type 2* sends a command (CMD frame as dotted line) to its I/O handler via UNIT *type 2*. For instance, we may use commands to set some features of I/O handlers or actually send commands to peripheral devices. This is of course possible — by declaring a CMD object in the corresponding UNIT we define the link. The Vfsm editor also allows us to use command names instead of numbers.

Properties

Commands are numbers. To make commands more understandable for human beings we use names instead of numbers in the specification. By default, the definition of command names is stored in H- and IOD-files generated by Build. The IOD-file is then used by the RTDB as a source of command names. But command names may be defined in any H- and IOD-file, which is important for state machines that use several* CMD objects (several sets of commands). To distinguish between different command sets an object CMD has a property **Type,** which should contain the name (without extension) of the H- and IOD-files that contain the definitions of command names for that object.

A detailed description of object use and properties can be found in References 1 through 3.

An Interface to I/O Handler (UNIT)

We now arrive at objects of type UNIT. As for VFSM the **UNIT** type is a list of objects. The list defines objects that are used in a corresponding I/O handler or Output function. UNIT does not have any behavior; different UNIT types differ only in **properties**. The list of objects is completed by two properties:

- **Phys Address**: an integer number
- **Comm Port**: a string, e.g., "COM1"

These two properties are very hardware oriented. If necessary, they may be misused for passing any value to an I/O handler.

StateWORKS defines a few standard UNIT types. There is a group of UNITs that are used for accessing digital and analog inputs and outputs on the graphical user interface of the SWLab simulator: DI8, DO8, NI4, and NO4. There is also a StandardUnit, which defines a list of objects (Appendix H *IOD file of the StandardUnit* shows the IOD file of the UNIT). The StandardUnit is actually a kind of template that has been used to program a standard I/O handler able to work with I/O drivers in form of DLLs. The I/O handler is programmed in such a way that only the objects commented as "must-not-be-changed" are constant entries that must stay there. All other object types may be adjusted to the application: removed

---

* The topic of several CMD objects in a state machine is too specific for the book. Detailed information is included in the StateWORKS Studio help.

or expanded with additional objects. The command names and values (see the C section of the IOD file) can be also defined according to the applications requirements.

VFSM and UNIT contents are used in I/O handlers and output functions, which are the programmable parts of an RTDB-based application. Therefore, Build generates H-files that can be used by programs written in C or C++. The H-file contains the names used in the virtual environment (Input, Output, State, and CMD names).

A UNIT type may be replaced by a VFSM type if the *Phys Address* and *Comm Port* are not used. This especially makes sense for output functions that are closely related to a certain state machine and use the same objects.

When specifying the list of objects needed for programming of an I/O handler, we may, in principle, use any object types. Using a DI object for writing digital inputs to the RTDB or using DO to write digital outputs to an I/O handler is obvious. The issue may not always be so clear if the signals sent to output devices are messages including some complex information that must be assembled in the code of an I/O handler. Is then a command a good choice? It might be, but a change of a parameter or a string sent directly via a TAB object may be a better solution in other cases. The goal is to trigger in some way a code in the I/O handler, which processes the request and eventually transmits a message, information, command, or whatever name we like to use in a certain situation. Hence, the object used as a trigger element depends on the application.

This last reflection also explains the possibility of putting any code, as well as those of output functions, in I/O handlers. We need just a trigger point to access an application specific code, which completes the RTDB functionality. We can have the code in the I/O handler but using OFUN as a gate for user-written software has some advantages.

An Interface to a User-Written Function (OFUN)

RTDB objects store data and generate control values corresponding to the actual data value. At present there are several types of objects — they have been developed according to known requirements while designing control systems in several application domains, especially industrial control and telecommunication. Of course, the defined object types cannot cover all situations that may be encountered in the design of control systems. Therefore, the RTDB also contains an object **OFUN,** which is an interface for user-written functions. In other words the object of type OFUN is a substitute for a hypothetical object whose features are defined by a function.

OFUN is an input/output object; i.e., it is controlled by actions and it generates control values. An action is an integer number that is passed to the called function as a parameter. The control values are integer numbers — return values of the called function. By that arrangement OFUN object is able to pass to the function the requirements "what to do" and to receive from the function the result of the calculation or operation or just information about success or failure.

The use of OFUN offers a simple input/output object that assures an integration of an Output function as if it were a standard RTDB object. It means also that we may build up a set of functions that find application in several projects via corresponding OFUN objects.

Example

The state machine of type Pressure in the project Pumps (see *Example — Pressure supervision* in Chapter 8 and Appendix L *Pumps supervision project*) contains an object OFUN. For this object there is one action Ofun_CalcLimit (value 1) defined — the name describes clearly "what to do": calculate limit(s). The function returns values:

- 0: wrong owner (Unit Name — see Properties below)
- 1: limits calculated
- 2: the calculation failed (wrong parameter ≠ 1)

Although the two failures signaled by numbers 0 and 2 belong to specification errors and not to runtime errors, we use them in the example to demonstrate the discussed issue: we have specified two Input names: Ofun_OwnerError (0) and Ofun_ParameterError (2) on them and use the names to break the starting process by enforcing a transition from the state *Starting* to *Idle* and to generate a corresponding alarm if the limits are not calculated (in fact, it is no use for any pressure supervision if the limits of the SWIP objects are not set).

Appendix M *Output function CalcLimit()* shows the function Calc-Limit() used in StateWORKS project Pumps for specification of the state machine Pressure.

Properties

An object of the OFUN type has two properties:

- **Function Name**: the name of the called C/C++ function
- **Unit Name**: the name of the UNIT or VFSM object which defines (we say also "owns") the function

The Output function accesses RTDB objects. The OFUN defines the list of objects "seen" by the function in form of UNIT or VFSM.

A detailed description of object use and properties can be found in References 1 through 3.

Just XDA

Eventually, we discuss the last type of RTDB objects — the **XDA**. It is an unusual type: it does not have a truly obvious name, it is just XDA. Objects of type XDA find two completely different applications.

Memory for OFUN

An XDA object is a memory for OFUN. To fulfill this task XDA has a **property**:

■ **Size**: a number (integer >= 0), which declares the size of the reserved memory in bytes

When writing an Output function that contains a thread we need some memory. By specifying an XDA object with memory we may use it while programming the function.

Internal Value as a Control Value

First, XDA is an input/output object like OFUN or CMD, whose control values and actions are defined by the designer of the state machine. Therefore XDA could be used as a substitute for a CMD object to pass information (commands) from Master to Slaves. In this application the CMD object is superior to XDA because CMD values can be names "seen" by RTDB clients and XDA are just pure integers. Also the predefined Clear action of the CMD object makes the use of CMD more comfortable than XDA. To avoid any misuse of a CMD object it cannot be used as input and output in the same state machine: the only exception to this rule is the output value Clear whose use is well defined and very limited. In other words, the CMD type has been designed to function very well as a link between Master and Slave. Therefore, we would not recommend replacing it with XDA without any recognizable advantages. XDA objects find applications in communication between state machines and I/O handlers, where they are very useful to pass (acknowledgment) information from the I/O handler to state machines.

Because of its simplicity we would sometimes think that we could use XDA just to store some information (as a number); as opposed to the CMD object there are no formal restrictions that would not allow it.* More careful analysis of such situations shows inevitably that we are just trying to corrupt the state machine by storing some flags instead of introducing missing states. Any XDA that is used in the same state machine as input and output is suspicious — it is very probable that the design is not correct. An XDA object used in one state machine as an output object and in another state machine as an input object might work, supporting or completing commands in some way.

A detailed description of object use and properties can be found in References 1 through 3.

Recommended Reading

1. SW Software: StateWORKS. Reference Manual for the Class Library.
2. SW Software: StateWORKS Development Tools. User's Guide.
3. SW Software: StateWORKS Studio. Help.

---

* Actually there is one restriction: XDA must not be used as input (control value) and output (action) in the same state as its behavior becomes unpredictable.

Chapter 16

Debugging Vfsm

Testing a Vfsm Application

Any nontrivial software must be tested. In testing, errors are detected and corrected — this process is known in software development as debugging. A program may have two kinds of errors: coding and application failures. Obvious coding errors manifest themselves by crashing that program or other programs, and in the worst case the operating system. Several not so obvious coding errors result in misbehaving of the application, so "simulating" in a way logical errors. A notorious example of such a failure is the unfortunate comparison operator (= =) used in C/C++, which often stays undetected by a compiler if confused with the assignment operator (=). I would like to have the money corresponding to the time invested in searching for this kind of simple error. Application errors mean that the program does not realize the requirements. Hence, the reasons for application errors are very often not clear — they can be logical errors or coding failures.

Classical debugging is done in a program, which means that we are trying to find the erroneous code and correct it. Such debugging is not easy as the two aspects, code and application errors, are overlaid.

In StateWORKS the debugging is clearer — in principle we are looking only for application errors. The probability of a coding error is vanishingly low as the application is built using a very robust, standard code (RTDB), which has been used and tested over many years in diverse applications.

Thus, for the user, debugging of the RTDB-based application means testing whether the system of state machines works properly. Testing of a state machine may be tricky. To facilitate testing a few facilities must be available, such as a trace facility, a debugging mode, an automation of test sequences, and a service mode. Good, up-to-date documentation rounds out the development environment.

System Consistency

Debugging of the RTDB application begins with a start-up during which a **SULOG.TXT** file is produced. This file is a log file with a list of objects that could not be built due to some inconsistencies in the system specification file SWD. The inconsistencies are allowed on purpose in StateWORKS to test not-quite-complete applications. Of course, the ultimate application should produce an empty SULOG file.

An example of a SULOG.TXT file is shown in Text 16.1. In a rather cryptic form it contains three warnings signaling missing:

- Cmd names list
- String for an alarm text
- Output function

Text 16.1   Example of a SULOG.TXT File

```
VFSM System Startup Log
-----------------------
Config File: ...\SWSystem\Spec\SWSystem.swd
Startup Time: 14-Jan-04 09:29:18
W1    4    Not found enumeration in/or IOD-File: joystickdigital
X:JoystickDigital:Cmd
W4    27   String Resource not found:   IDS_AL_START_ERROR
Axes:Al:Motor_Start_Problem
W135 8    Not found User Defined OutFunction OfuCheckDeviceNumber
Allsens:OFun:CheckDeviceNumber
There are:
   3 Warning(s)
```

The number in the second column (4, 27, 8) is an error/warning type — in all there are 30 types of them.

Trace

Trace means the ability to record all or selected steps performed by a program. All RTDB objects have the trace facility, which may be activated or disabled. Normally, the trace facility is disabled. If activated, the trace facility causes all changes of object states (and for some objects also the data values) to be logged into a **TRACE.TXT** file. Trace can be activated from Monitors by setting the Trace flag of an object to true (1). Any number of objects may be traced simultaneously. In addition, the TRACE.TXT file can be closed and opened again at any time from Monitors. After opening, the TRACE file is always empty. An example of a trace is shown in Text 16.2. A line in a trace file shows:

- The time of the event
- The object type
- The object name
- The numerical value of the object state
- The name of the object state (if available)
- The data value (if appropriate)

Text 16.2   Example of a TRACE.TXT File

```
VFSM System Trace File
----------------------
(Date: 18-Oct-04 21:57:22)
21:57:29 CMD   Main:Cmd                                    0
21:57:29 CMD   Pressure1:MyCmd                             1 Cmd_Start
21:57:29 VFSM  Pressure:01                                 3 Starting
21:57:29 CMD   Pressure1:MyCmd                             0
21:57:29 TI    Pressure1:Ti:Timer                          3 RUN
21:57:29 NO    Pressure1:No:SetPressure                    3 CHANGED  900
21:57:29 NO    Pressure1:No:SetPressure                    5 SET      900
21:57:29 OFUN  Pressure1:Ofun:ActualPressure_CalcLimit     1
21:57:35 SWIP  Pressure1:Swip:ActualPressure_Supervison    3 IN
21:57:35 TI    Pressure1:Ti:Timer                          2 STOP
21:57:35 VFSM  Pressure:01                                 4 Regulating
```

For instance, at 21:57:29 the object NO with the name Pressure1:No:Set-Pressure changed its state to SET (5) because its value had been set to 900.

The trace illustrates well the events in the state machine Pressure1: it received the Cmd_Start and went into the state *Starting* (3) where the following Entry Actions were performed: the Cmd was canceled, the Timer started, the Numerical Output set to 900 (two entries because the NO data was first changed and then set as output), and the Output function was called to calculate the switchpoint limits — the operation was successful, which was signaled by the return value 1. At 6 seconds later (see time stamp) SWIP object had signaled with its IN state that the input pressure was within limits, which forced the state machine to go to the state *Regulating*. Before it left the state *Starting* it had stopped the Timer.

The trace "description" corresponds of course exactly to the state machine specification — see Figure 9.17.

Debugging Mode (VFSM)

Probably the most difficult requirement is to test state machines "slowly," step by step. A state machine or even worse a system of state machines may have loops. A loop in a state machine means that a state machine performs several state changes or Input Actions in response to a single stimulus (condition change), and that there is a design error. The changes are too fast to be noticed by a human being. For a coded implementation we debug the code using a debugger **step mode**. A similar possibility is available in the StateWORKS development environment.

A **step** in the StateWORKS execution environment means to perform one state transition (with appropriate actions) or Input Actions specified in a present state. The use of the step mode is especially easy in SWMon where a state machine has three radio buttons to control the step mode: Run, Hold, and Step. Of course, we can control the step mode from other monitors by setting the RMo (Run Mode) and NSt (Next Step) attributes.

If the Run button is chosen (default) as in Figure 16.1 the state machine runs and the Next box to the right to the radio buttons displays "-/none": no Input Action and no transition.

Marking the Hold button as in Figure 16.2 stops the execution of the Vfsm Executor but the next state is displayed in the frame right to the buttons. In the example below the next state is *Starting*.

Clicking on the Step button forces the Vfsm Executor to perform one step — in the example the transition to the state *Starting* (Figure 16.3

Figure 16.1   SWMon: VFSM Pressure1 is in the Run mode.

Figure 16.2   SWMon: VFSM Pressure is in the Hold mode.

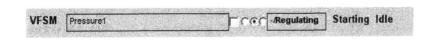

Figure 16.3 SWMon: VFSM Pressure1 has executed one step.

| VFSM | Pressure1 | | ⌐ ⌐ ⊙ ⌐ A/Idle | Starting Idle |

Figure 16.4 SWMon: VFSM Pressure1 has executed one step and the timer expired.

| ALARM | | AL | | STAYING | Pressure1: Pressure regulating error (83.082mB 1 |
| VFSM | Pressure1 | | ⌐ ⌐ ⊙ ⌐ -/none | Idle Starting Idle |

Figure 16.5 SWMon: VFSM Pressure1 has executed the second step.

shows the SWMon display after the step). The system returns to the Hold mode and waits for the next step. In the example the next step means a transition to the state *Regulating*. (Note that the actual state and some previous states are shown by a rightwards scrolling of the right-most field.)

If we wait until the timer expires, the box Next changes and signals now that there are Input Actions and a transition to the state *Idle* to be done (Figure 16.4). We see that all changes of the inputs are observed in the Hold mode — the next actions or transition displayed corresponds always to the actual input situation.

After the second step the Alarm is generated (it was the announced Input Action) and the state machine Pressure1 goes to the state *Idle*. There are no actions or transitions due in this moment (see Figure 16.5).

If there are no Input Actions or transitions due when we click on the Step button the button stays marked and the Vfsm Executor will perform the appropriate actions or transition if the corresponding condition becomes true.

Command Files

Testing of the application is normally a long process. Using StateWORKS Studio we are testing from the project's beginning. Testing means a repetition of different scenarios. This procedure is tiresome and requires automation. The SWTerm monitor is used for generating automated test sequences. While running, SWTerm generates a log file **SWTerm.log**, which stores all typed sequences. An example is shown in Text 16.3. Renaming the log file to **SWTerm.cmd** changes it to a **command file**. Of course, the command file can be also prepared and edited in any text editor. Other file names may be used for command files if we need to have several of them. Starting the SWTerm monitor with the argument "-cSWTerm.cmd" instructs the program to open the command file and realize the command in the file line by line. Each line has to be acknowledged by the user.

The meanings of the commands used are:

■sw c	Connect to RTDB
■sw n IL.	Get the path of the (application) SWD file
■sw o CMD	Display names of CMD objects
■sw s Main:Cmd.PeV 1	Set the 1 value (Cmd_Start) to Main:Cmd.PeV
■sw g Main.StN	Get the State name of the state machine Main
■sw g Pressure1:Ni:ActualPressure	Get the value of Pressure1:Ni:ActualPressure input
■sw s Main:Cmd.PeV 2	Set the 2 value (Cmd_Stop) to Main:Cmd.PeV
■sw d	Disconnect

Executing the command file as in the example in Text 16.3 will produce the command and answers as shown in Text 16.4.

Text 16.3 The SWTerm Command File

```
sw  c
sw  n  IL.
sw  o  CMD
sw  s  Main:Cmd.PeV  1
sw  g  Main.StN
sw  g  Pressure1:Ni:ActualPressure
sw  s  Main:Cmd.PeV  2
sw  g  Main.StN
sw  d
```

Text 16.4   The monitor SWTerm executing the command file

```
Taking commands from a command file SWTerm.cmd
to continue press Enter
q to quit program
h to display help
v to display the version

Enter command: sw c
Using default Host address(LOCALHOST) and Port number(9091)
CONNECTED

Enter command: sw n IL.
...\Projects\Examples\Pumps\Conf\Pumps.swd

Enter command: sw o CMD
CX1000:Cmd:01
Device:MyCmd
Main:Cmd
Pressure1:MyCmd
Pressure2:MyCmd

Enter command: sw s Main:Cmd.PeV 1
Value set

Enter command: sw g Main.StN
StN = On

Enter command: sw g Pressure1:Ni:ActualPressure
Dat = 88.2548

Enter command: sw s Main:Cmd.PeV 2
Value set

Enter command: sw g Main.StN
StN = Idle

Enter command: sw d
DISCONNECTED

Enter command:
```

Service Mode

While testing we like to have a possibility for simulating values which
influence a control. This feature called **Service** or **Force mode** is especially important for external input signals. StateWORKS provides a Service

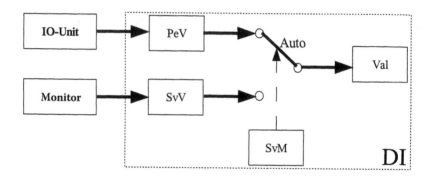

Figure 16.6 Service mode for a DI object.

mode for all external I/O objects: DI, DO, NI (represented by SWIP), and NO. Additionally the internal I/O objects CMD and VFSM can also run in service mode allowing us to debug the Master–Slave interface. Note that this possibility is very useful for testing parts of a project which we have not completed.

The service mode for DI is obvious — it allows us to test the system without its hardware digital inputs. The default mode is the Auto mode which means the DI values are coming from the I/O handler, effectively from the hardware. The Peripheral Value (PeV) in Figure 16.6 comes from the hardware via the I/O handler and is passed in Auto mode (SvM = false) to Val stored in the RTDB. In Service Mode (SvM = true) Val gets the Service Value (SvV) set by the Monitor.

The CMD object used as interface between a Master and a Slave state machine belongs to the Slave (CMD-IN). The service mode for CMD (see Figure 16.7) allows switching the command value between the value set by Master (CMD-OUT) in Auto mode and Service Value (SvV) in service mode set in Monitor.

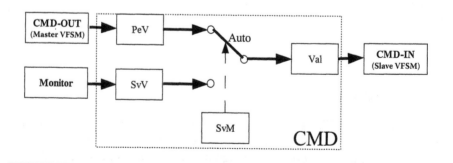

Figure 16.7 Service mode for a CMD object.

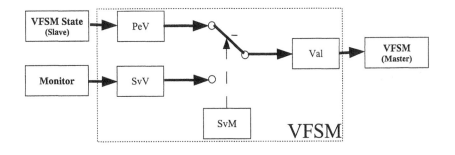

Figure 16.8   Service mode for a VFSM object.

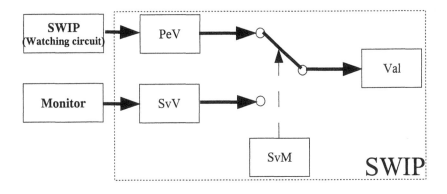

Figure 16.9   Service mode for a SWIP object.

The service mode for VFSM (see Figure 16.8) means simulating the Slave state for a Master. In the Auto mode the Master gets the Slave state. In service mode Master gets the service value set in Monitor.

In Auto mode the SWIP Val (see Figure 16.9) comes from the supervising part which compares the SWIP limits with the present value of the supervised object. In service mode the value is set in Monitor. Used for NI object the SWIP service mode is effectively a replacement for the NI service mode (we are less interested in the NI data value than in the control value determined by SWIP which supervises the NI object).

For completeness, StateWORKS provides also a service mode for DO and NO outputs. This feature is useful for testing the hardware, at least to be sure that the cabling is correct. In Auto mode the hardware receives via I/O handler the DO value as stored in the RTDB (see Figure 16.10). In service mode the value passed to I/O handler is set in Monitor.

The service mode for an NO object is rather crude — it allows switching on and off the NO output. In such a way we may pass the numerical output value by hand.

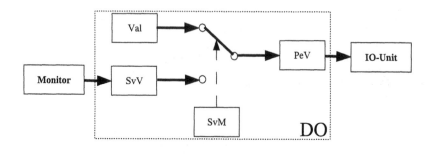

Figure 16.10   Service mode for a DO object.

The Role of Documentation

The software documentation is the only information source for several groups within the company producing the product. Several tasks must be performed in parallel to programming:

■ Planning the testing of a design requires a good overview of the software.
■ The product test team must create the application test plans.
■ Customer documentation must be written.
■ Product support and maintenance needs detailed information.
■ Further product development should not count on information stored as code.

Because of this, complete and up-to-date software documentation is a key requirement for all software manufacturers. The creation of this documentation has to start as soon as possible to ensure that all other tasks like test plans or customer documentation will be done in time. During debugging several things will change: state machines, the object properties, the command files. These changes influence other software documents produced by cooperating teams.

The entire content of the StateWORKS application — the configuration of a system of state machines, state transition diagrams, state transition tables — are available as JPG or WMF files. All this information is also produced as XML documents.

Because in StateWORKS any change in state machine behavior can be done only "officially" in the StateWORKS Studio I.D.E., the documentation is always up to date. There is no way to corrupt the documentation as it is the inherent part of the development system reflecting always the actual state of the developed application. In short, in StateWORKS the distance to the updated documentation is very short — just one click.

Chapter 17

What Is StateWORKS?

Compared with Specification Methods

Specification methods are code oriented; i.e., they try to specify the software. The entry point is of course the requirements of an application, but the specification takes into account the implementation. In rare cases, in companies that can afford it, there are two specification levels: the first abstract specification (it is something like a formalization of the customer requirements), which is actually the true application specification. The second specification is then an implementation (software) specific specification.

The specification is considered an initial phase, which is good for learning the requirements of a project and exploring the design possibilities. It is not treated seriously afterwards as it does not count — only the code counts as that is what really works. Specification is by definition not complete, as at the beginning of the project many details are not known and could be only defined later. Specification is also not reliable as several difficult details are intentionally omitted assuming that they will be solved during programming.

When programming starts, the specification begins to lose its validity as it is not actualized: changes introduced during programming stay buried in the code. The value of the specification as documentation is minimal as it can not be trusted.

Working with StateWORKS the specification result is *the* result — there is no programming phase that can compensate for omissions and failure of the specification. In that case the specification *must* be complete. StateWORKS specifies the behavior of an application assuming that the control flow (behavior) is performed by a standard RTDB-based code.

Compared with Agile Methods

Agile methods do without any detailed specification, assuming that it is impossible to specify fully an application. Thus, they assume that the code will be the specification and it will be complete if it works. It is true but has one weak point: the specification language. Considering a programming language as a specification language is a mistake, as no one later understands the specification. The other weakness of the programming language considered as a specification language is that behind the language there is no model that may help us interpret the lines of code. Without such a model it is impossible to translate (mentally) a text in the form of a source code into a comprehensible picture describing the behavior of software (the application). The other interpretation of an Agile approach could be just a plain statement: software does not need any specification; code is all — then there is no basis for any discussion.

StateWORKS avoids much intricate coding, leaving only the data acquisition parts and data processing procedures. The main software path — the control path is completely specified and the "specification language" is positive logic algebra in the virtual environment. The model used for behavior specification is a state machine model, which is an established and proven means known and used over several years. Using such a model we have a chance to produce a comprehensible specification that can be also verified as it is based on formal mathematical principles.

In a similar fashion as for Agile methods, using StateWORKS we start the implementation immediately, but on a higher abstraction level: in StateWORKS the implementation means generating a specification, in Agile methods — coding.

Application Areas

StateWORKS concepts have been applied to a wide variety of projects, over several years. The results were discussed by Wagner et al.[1–3] These have ranged from industrial controls and specialized measurement systems to telecommunications switching and protocol handling. The projects have not been trivial: the first implementation of the Vfsm principle was for a semiconductor production line, using several mini-computers, in 1988–1990. It was in wide use at AT&T and later Lucent for their international telecommunications switching products and has been shown to reduce development cost and also improve quality (the results are presented by Flora-Holmquist et al.[4]). The application of the Vfsm concept for modeling technique is discussed by Mahoney et al.[6]

Because the StateWORKS tools impose a certain structure on the software, it has sometimes been possible to successfully and rapidly complete a project that had been thought to be impossible, on account of its apparent complexity. If a complex project is hard to comprehend, then coding in the classical way, with any available language, will not make it easier to understand, and failure can result.

The full StateWORKS system with the RTDB has been in use since 1977, and has been very successful wherever applied. StateWORKS can be used either as a specification tool or preferably as a complete development and execution environment. Although StateWORKS can be used in development of any application with nontrivial control requirements, some domains seem especially suitable for StateWORKS use. Embedded systems are here the primary example, especially as a replacement for Programmable Logic Controllers (PLC).

Due to the specific hardware used, PLC systems were for a long time isolated from the mainstream of software development. Nowadays, robust and reliable PCs are replacing PLC hardware in many industrial applications. Moving the old-fashioned PLC programming onto a PC is a consequence of programming methods used in the PLC world: the use of simple interactive programming tools without a typical burden of compiling, linking, building, etc. plus very conformable debugging facilities. As we discussed it in Part I (see the section *PLC* in Chapter 1 and the *Hardness of software* section in Chapter 2) in spite of some nice features of the PLC tools, the methods are hopelessly old-fashioned and reached their limits some time ago. The attempts to improve them (such as the IEC-61131-3 standard) are a kind of patch, which does not change the overall bad programming style. StateWORKS could be a good replacement in that domain: it is not only a modern and effective development tool but also possesses nice features, such as no software building problem, a very comfortable debugging environment, and automated generation of documentation. Wagner[5] discusses the topic in detail.

Several aspects of Vfsm method, its application, and related topics are discussed in case studies and technical notes available on the Web.[7]

Recommended Reading

1. Wagner, F., Wolstenholme, P., "A modern real-time software design tool: applying lessons from Leo," *IEE Computing & Control Engineering* (February 2003).
2. Wagner, F., Wolstenholme, P., "Modeling and building reliable and re-usable software," *Proceedings of the ECBS'03*, Hunstville, AL, April 2003.

3. Wagner, F., Wolstenholme, P., Wagner, T., "Closing the gap between software modeling and code," *Proceedings of the ECBS'04*, Brno, May 2004.

4. Flora-Holmquist, A. R., Morton, E., O'Grady, M. G., Staskauskas, M. G., "The virtual finite state design and implementation paradigm," *Bell Labs Technical Journal* (1997): 97–113.

5. Wagner, F., "Going beyond the limitations of IEC 61131-3," SW Software Technical note, 2005.

6. Mahoney, M., Tzilla, E., "Modeling platform specific attributes of a system as crosscutting concerns using aspect-oriented statecharts and virtual finite state machines," paper presented at *6th International Workshop on Aspect-Oriented Modeling*, March 14, 2005, Chicago.

7. http://www.stateworks.com.

Appendix A

Case Studies

The www.stateworks.com Web site contains a few case studies:

- Traffic Light Control
- Microwave Oven Control
- Gas Control
- Industrial Control
- Dining Philosophers Problem

We have included three of them in the following three appendices; material from the other two was partly used in the book. The text of the included case studies has been slightly changed and adapted to the requirements and style of the book.

The Web site also contains several technical notes: at the time of writing there were the following topics represented:

- The Virtual Environment and Positive Logic Algebra
- What Is StateWORKS?
- Hierarchical systems of state machines
- New version of CMD object (Commands)
- String Object (STR)
- How to Write a GUI for StateWORKS Applications
- Standard Interface for StateWORKS Standard Executor
- Debugging state machines
- A flowchart is not a state machine

- Moore or Mealy model?
- StateWORKS — specifying control software instead of coding
- StateWORKS IO-Unit for Velleman K8055 (or VM110) boards
- Completeness of information in the virtual environment
- Going beyond the limitations of IEC 61131-3

The last technical note is included as Appendix E.

Appendix B

Microwave Oven Control — Use of StateWORKS Development Tools

Topic

The control of a microwave oven is relatively simple and does not present any challenge to the designer. We have chosen it to show the usage of real-time database (RTDB) objects. The RTDB objects are of various types and may realize some control functions. Correct use of these functions simplifies the state machines. Some obvious examples of control functions are counters, which count commands, events, clock pulses (timers), etc. In this example we want to show the usage of switchpoints (Swip), which in RTDB-based systems are used to supervise whether an object value stays within limits.

The requirements are:

> The oven has a Run pushbutton to start (apply the power) and a Timer that determines the cooking length. Cooking can be interrupted at any time by opening the oven Door. After closing the Door the cooking is continued. Cooking is terminated when the Timer elapses. When the Door is open a Lamp inside the oven is switched on; when the Door is closed the Lamp is off.

The control system has the following inputs:

- **Run** push button — when activated starts cooking
- **Timer** — while this runs keep on cooking
- **Door** sensor — can be HIGH (door closed) or LOW (door open)

and the following outputs:

- **Power** — can be HIGH (power on) or LOW (power off)
- **Lamp** — can be HIGH (lamp on) or LOW (lamp off)

The knobs to set the power and time-out values are irrelevant for the control state machine. The behavior of the microwave oven control is determined by the Run push button, Timer, and Door sensor.

First Simple Solution

The first approach is simple and we consider only the basic requirements as listed above. We later analyze the missing control function and specify a more complete control system.

The state transition diagram for the simple solution is shown in Figure B.1. The details of the state transition diagram are in the ST table. To display it you need to use the program StateWORKS Studio. For the purposes of this study we show here only the table for the state *Idle* (Figure B.2).

This simple solution has some weak points. Of course, microwave ovens are manufactured in different flavors with slightly different controls. Thus, we cannot analyze all possible variants we see on the market. For the purpose of our exercise we will consider the timer. The timer usage

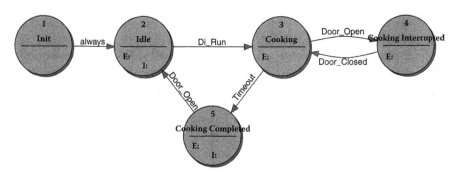

Figure B.1   First Microwave oven control: the ST diagram.

Entering the state the switchpoint is activated.
Opening and closing the door switches the lamp on and off.
If the Run signal becomes active and the Timeout value is not zero the state machine goes to the state Cooking.

Idle	Entry action	Swip_Timeout_On
	eXit action	
	Door_Closed	Do_LampOff
	Door_Open	Do_LampOn
Cooking	Di_Run	

Figure B.2   First Microwave oven control: the ST table of the state *Idle*.

is not perfect; namely, the control system always starts, even if the time-out value is set to 0. To reset the system for the next start we have to open and then close the door: if you do not like this, then you will find the exercise of changing the design very educational.

More Realistic Control

To achieve a more elaborate control as shown in Figure B.3 we introduce a condition `Swip_TimeoutNotZero`, which together with `Di_Run` (using logical AND operation) forms the condition for the transition from the state *Idle* to *Cooking* (see Figure B.4). A Switchpoint in the RTDB is a mechanism for testing any numeric value, and producing a set of Control Values in the positive-logic convention employed by StateWORKS to govern possible

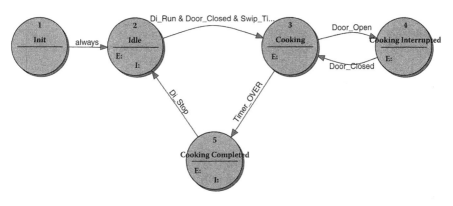

Figure B.3   Microwave oven control: the ST diagram.

Entering the state the switchpoint is activated.
Opening and closing the door switches the Lamp on and off.
If the Run button becomes active and the Timeout value is not zero the state machine goes to the state Cooking.

Idle	Entry action	Swip_Timeout_On
	eXit action	
	Door_Closed	Do_LampOff
	Door_Open	Do_LampOn
Cooking	Di_Run & Door_Closed & Swip_TimeoutNotZero	

Figure B.4   Microwave oven control: the ST table of the state *Idle*.

transitions. Instead of using the Control Values directly we use control names (see more detailed explanation later in "RTDB object" section). Commonly, we employ names such as In_Permitted_Range, Too_High, and Too_Low.

The specification is abstract, so we do not care at this moment how we might get the conditions. At this moment, the Vfsm specification defines a link to the real object in the RTDB by choosing the needed object types and using their control values to define control names, as, for instance, Swip_TimeoutNotZero.

Later, during the detailed system configuration, we can decide which specific timer will be used for oven control and link its time-out value with a parameter (set by a timer knob on the microwave oven front panel). That parameter, in turn, will be an object to be supervised by a switchpoint. We may show the dependencies specified in the RTDB by the diagram in Figure B.5.

RTDB Objects

The RTDB consists of objects. Objects are of different types; i.e., they have different properties; specifically they have different Data and Control Values.

A parameter object may have Data of type integer, float, string, etc. It may also have a Control Value but we do not discuss it here as it is irrelevant to the example. In our example Parameter stores the time-out value, which is an integer representing the number of seconds.

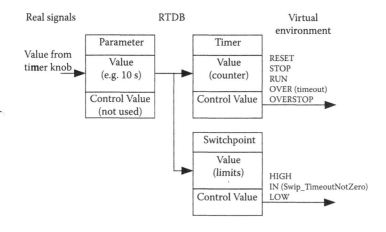

Figure B.5 Microwave oven control: dependencies among RTDB objects.

The timer data value is its counter output, which represents the elapsed time from the timer start. The timer object compares continuously the counter output with the time-out value (in this case supplied by the Parameter object) and defines the timer Control Value. If the timer is not started yet the Control Value is RESET. If the timer runs (the counter counts some time pulses) the Control Value is RUN. If the counter output equals the time-out value the Control Value is OVER.

A switchpoint object (SWIP) compares some Value with its Data Values, which represent in this case Low and High levels according to the diagram in Figure B.6.

In our example the Switchpoint object compares its Data (1 and a very large number) with a Parameter value which is a time-out. The result

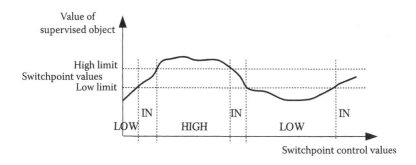

Figure B.6 Microwave oven control: the switchpoint (SWIP) object generates the control value.

of this comparison is Switchpoint Control Value which may be LOW, IN, or HIGH. As we are interested only in the value 0 of the Parameter we use the Control Value = IN as a condition name `Swip_TimeoutNotZero`. Note that the IN range includes the two limit values.

Yet Another Change

To demonstrate the flexibility of the RTDB we explain how to change the time-out value if it is determined by a potentiometer. The potentiometer delivers an analog signal — a voltage. So, we use a numerical input (NI) object in the RTDB to store the voltage and then use this object as a source of the Const value for the Timer object and as an Input for the Switchpoint object. In other words, the only change is to replace the Parameter with the Numeric Input.

Conclusions

A state machine determines the behavior of a control system. The complexity of the state machine depends, among other factors, on the means that are to be used to build the control system: the input/output system (hardware interface) and the system resources (timers, counters, etc.).

The RTDB provides a set of already-prepared objects, which can simplify the design of a control system by implementing some general control functions. Here we have shown one of these control functions: supervision of a certain value.

You may have noticed that the requirements as specified in the beginning were not very detailed. We did that on purpose as it is often the situation with which we are confronted in a real project. Not until we see the first solution do we "discover" that it is not what we have expected. Well, formally the first simple solution fulfilled the requirements. To start a project with an incomplete specification is not desirable, but it is a common practice and often unavoidable. Anyway, eventually we completed revision of the requirements and the second solution seems to be more realistic.

A real project may require additional control functions. A microwave oven usually has a rotating platform, which turns when the power is applied: separate controls for the power and the motor may be required. There are microwave ovens that "store" the Start signal even when the time-out is 0: in such a case we may first push the Start button and later, at any time, we start cooking by setting the timer. Another problem may be the setting of the time-out value: it may be a direct digital signal, an input from a keyboard, or an analog setting adjusted by a potentiometer: these possibilities require different solutions in the RTDB. However the details may vary, any requirements can be transformed to a neat Vfsm specification implemented by means of the StateWORKS system.

We have used such very simple examples in order to illustrate some of the important aspects of StateWORKS: the usage of RTDB objects. You may play with this example using the StateWORKS simulator (SWLab). This simulation tool, which supports the development of a control system, can be downloaded from our site. Using StateWORKS Studio you may change the behavior of the MWOven control and test it with SWLab. SWLab simulates inputs and outputs and contains the Vfsm Executor. To see what is going on, do not merely open the state machine and its states, but investigate the "Dictionary" and "Name" pull-down menus from the top tool-bar, as well as the open "Project" window to see how all the objects are defined and configured.

Demo

The MWOven project can be downloaded for test. You may inspect and change the project using the StateWORKS Studio. You may run and test the MWOven application using SWlab and SWMon or SWTerm.

Test: start SWLab and open the MWOven.swd file. SWLab displays then the DI inputs: Door and Run, DO outputs: Power and Lamp, and the NI input: CookingTime. As the CookingTime is initialized to 2048 you would probably like to set it to some lower value or 0 in the beginning. (Take no account of the scale markings, which are only intended as a logo for a numeric setting. You can quickly alter the setting by moving the pointer with a mouse select/move operation.) Alternatively, you may set the Offset property of the MW:Ni:CookingTime when specifying the RTDB objects to -2048.

Appendix C

Gas Control — Hierarchical System of State Machines

Topic

The Gas example is taken from our User Manual. It has been modified for the purpose of this case study. The changes allow the system to be tested with SWLab, without needing SWMon although using a monitor makes testing more comfortable.

The example is of a control system to control gas inlets of a vacuum chamber used by semiconductor manufacturers (Figure C.1 shows the control elements). The system contains flow control and pressure control elements. Three flow regulators supply three different gases to the chamber and they are controlled by a state machines Flow. The chamber is used for a manufacturing process that requires a certain vacuum in the chamber. The low pressure in the chamber is produced by vacuum pumps, not shown on the diagram. Effectively, the vacuum in the chamber is determined by the pumps and the gas flow. The pressure is continuously monitored and if it exceeds the required range the process is interrupted and the gas flow must be discontinued.

The system as designed contains five state machines (Figure C.2): three state machines Flow for gas flow control, the state machine Press to monitor the vacuum in the chamber, and a state machine Gas that is a Master, which coordinates the activities of the state machines Flow and Press.

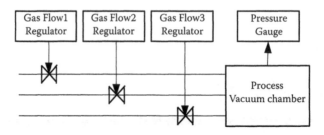

Figure C.1   Gas control.

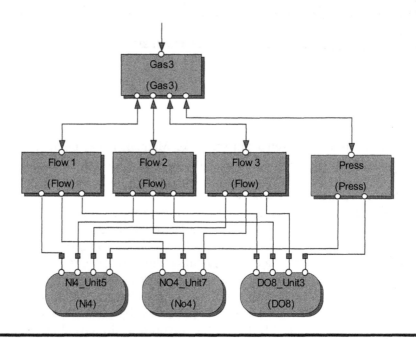

Figure C.2   Gas: the SMS diagram.

Flow Control

A gas flow regulator regulates the amount of gas passing through the gas inlet. The state machine Flow sets the required value of gas flow and monitors the true value of the flow. It is a kind of supervisor of the gas flow regulator. The gas flow regulator is a PID regulator and corrects the fast flow deviation. The Flow control reacts to permanent flow deviations, which cannot be corrected by the gas flow regulator, issuing alarms and closing the flow in emergency cases.

The state machine Flow accepts commands (numbers) and sets an analog signal `Flow_Ao` (set value for the gas flow regulator) and a digital signal `Flow_Do` (open/close valve). As a feedback from the attenuator the state machine receives the actual gas flow values as an analog signal `Flow_Ai` and a digital signal `Flow_Di` indicating the regulator position (open / closed). Three commands determine regulator operations:

- `Open (1)`: opens the gas flow by setting the High value of the digital output Flow_Do
- `Close (2)`: closes the gas flow by setting the Low value of the digital output Flow_Do
- `Regulate (3)`: sets the gas flow to the value determined by the numerical output Flow_Ao

If the command `Open` has been carried out the state machine should check whether the gas valve has been opened. If the valve does not open after a certain time an alarm should be issued.

If the command `Regulate` has been carried out, the state machine should check whether the gas flow has reached the required value. If the flow cannot reach the value in certain time an alarm should be issued.

The actual gas flow is measured and its value delivered as an analog input signal `Flow_Ai` to the control system. Normally, it is required that the flow value stay within certain limits. If the flow value exceeds the required range, the state machine should after some delay issue an alarm and count this event. If it happens more than a certain number the state machine should issue immediately the alarm if the flow value exceeds the range.

You can find the details of the state machine Flow in its specification, especially in the ST tables. We show here only the state transition diagram in Figure C.3, which gives a general impression about the Flow control.

Note that the state machine Flow has two Do outputs. The output DoValve is a control relevant output used to open/close the gas flow valve. The output DoRegulating is only for the purpose of this case study to show in SWLab that the gas flow value has reached the required value (i.e., the state machine Flow is in the state *Regulating*).

Monitoring the Pressure

The state machine Press does not perform any true control–it does not have any output like Do or No. It rather monitors the pressure and represents the result to the Master with its state. Hence, the role of this kind of state machine is to isolate the Master from measurement details

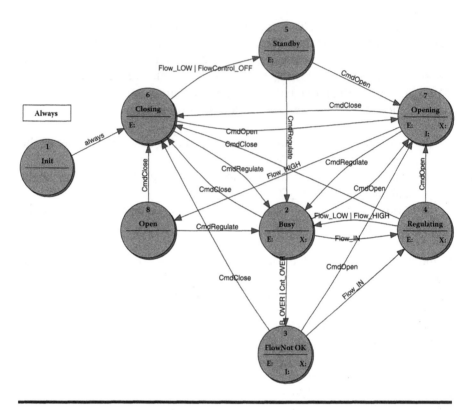

Figure C.3   Flow: the ST diagram.

and to supply the Master more abstract but truly relevant control information, like the pressure is ok or wrong.

The actual pressure value delivered by the numerical input Ai is controlled by a switchpoint SwipPress and by a Timer. If the measurement is on, the timer is started. If the timer expires before the pressure (Ai) reaches the required range Press signals the bad pressure with its state and issues an alarm. If the pressure returns to the required value, Press signals it with its state. Thus, during measurement Press changes among states: *Busy*, *PressOK*, and *BadPress* signaling to the Master the pressure value.

Note that similar to the state machine Flow we added an output Do, which is irrelevant for the control but is used to switch on a lamp in SWlab to show that the gas pressure is ok.

The state machine Press is shown in Figure C.4. The ST tables contain the details.

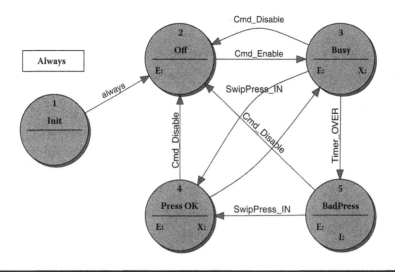

Figure C.4   Press: the ST diagram.

Gas Control

The state machine Gas has four slaves: one Press and three state machines Flow. The slaves get commands from their Master — the state machine Gas. The state machine Gas uses the states of its slave to supervise their activity.

Normally, the state machine Gas would receive a command as its primary input. As we cannot send a command from the SWLab we replace for the purpose of the case study the Gas command by a Di: HIGH value means command On, LOW value — command Off. The correspondence is not 100% but it is sufficient for the example.

Considering their behavior the three state machines Flow are the same. Therefore, we have only one type of state machine Flow in the system. The three state machines Flow controlled by the state machine Gas are three incarnations of the same state machine. The differences among the state machines Flow are in their properties such as time-outs, retries, flow set values, supervised flow limits, etc. These are just set as properties of state machine Flow objects. Note that the Flow machine does not regulate the flow: it merely passes an assigned parameter to the PID controller to fix the set point. This would be exactly the same were the PID controller to be a software package running under the same processor, rather than an external device: the state machine Flow neither knows nor cares about such details.

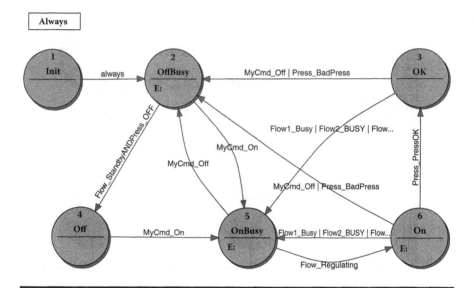

Figure C.5 Gas3: the ST diagram.

The state transition diagram in Figure C.5 shows that the behavior of the state machine Gas is relatively simple as the detailed control problems on the device level are handled by Press and Flow state machines.

The details of Gas control can be seen in its specification, especially in the ST tables. Note that the Input names in the state machine Gas are defined as complex expressions. This feature of StateWORKS makes the state machine specification very comprehensible, hiding the complexity of the transition conditions behind expressive names. For instance:

- Flow_StandbyANDPress_Off means: "all state machines Flow are in the *Standby* state AND state machine Press is in the *Off* state" and replaces the following logical expression:

  Press_Off AND Flow1_Standby AND Flow2_Standby AND Flow3_Standby.

- Flow1_BUSY means: "state machine Flow1 is in the *Busy* or *FlowNotOk* state" and replaces the following logical expression:

  Flow1_Busy OR Flow1_FlowNotOk.

Conclusions

The Gas control system with its five state machines is a typical example of a hierarchical system of state machines. The Master (Gas) and its Slaves (3xFlow and Press) communicate using a command–state interface: the

Master sends commands and supervises the reaction of Slaves to its commands by monitoring their states.

The case study also illustrates how such a system deals with numerical data. Although we have taken pains to point out that real data, particularly numerical data, is kept out of the state machine specifications generated when designing with StateWORKS, the control system does, of course, need to handle such data, such as pressure and flow settings.

Demo

You find the Gas project in the Samples dictionary (there are few variants of it there; the discussion in this case study applies to the Gas3 project) if you install StateWORKS Studio. Look at the state machines and system specifications using StateWORKS Studio and run it using SWLab and SWMon or SWTerm.

If you set Gas:Di:Cmd HIGH (switch to left) Gas state machine goes to *OnBusy* state and sends to all Flow state machines the command Regulate. The input values of the measurement instruments are simulated by Ni. The correct, expected values are:

- Press:Ni:ActualPressureValue in the range 322–415
- FlowN:Ni:ActualFlowValue in the range 1000–1200

The values result from the (arbitrary chosen) properties (scale factor, etc.) for these elements. The values and properties for all Flow objects are the same.

Do lamps signal if the set values (states) are achieved. The values of outputs No for Flow valves will be then set to 750.

Appendix D

Dining Philosophers Problem

Example

There are some variants of this problem. One of them reads:

> There are five philosophers sitting at a round table who do nothing but think and eat. Between each philosopher there is a single fork. In order to eat, a philosopher must have both forks.

This problem is used to discuss multi-process synchronization problems, like deadlocks and starvation. You get these problems if you put some restrictions on the way a philosopher grabs for a fork, for instance, first on the right and then waits for the fork on the left. But here I do not want to model synchronization problems — so I realize the working system where a philosopher starts eating if both forks at his sides are free. As this example has been invented as a Christmas gift to study starvation would not be a proper topic.

Actually, I took the wording from:

http://www.codeproject.com/csharp/FSMdotNet.asp?target=
state%7Cmachine&select=691061&df=100&forumid=
29430&fr=16.5#xx691061xx

where I took part in a discussion about state machines. You may find it interesting to compare the effort you need if you code something and use a ready-made execution system like StateWORKS.

A philosopher's behavior is simulated by a state machine represented by the state transition diagram in Figure D.1.

The eating time and thinking time are defined by separate timers. The forks are represented by XDA objects.

To simulate the problem we need then five state machines, one for each philosopher (see Figure D.2). The state machines are not a system of state machines; they are just five separate state machines. The dependencies among philosophers (state machines) come into being as they

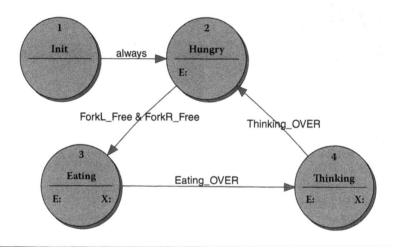

Figure D.1   Dining Philosophers: the ST diagram.

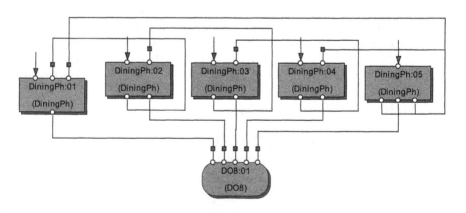

Figure D.2   Dining Philosophers: the SMS diagram.

use common forks (a left fork of one philosopher is the right fork of the neighboring philosopher).

Running the Example

The DiningPhilosophers example is available for download on our Web site. You may try it if you possess StateWORKS development tools. You may run the SWLab with the DiningPhilosophers example and monitor the system using SWMon, SWQuick, or SWTerm. The system uses digital outputs to indicate thinking philosophers: if a Do is *on* the philosopher thinks; otherwise he eats or is hungry. In SWMon you may change the time-out values for thinking and eating but I can assure you that for any combination of time-out values no philosopher will starve (well, assuming that you do not use very large values).

Appendix E

Going Beyond the Limitations of IEC 61131-3

Introduction

We start with an analysis of an example taken from the IEC document,[1] which specifies the 61131-3 standard for programming PLC (Programmable Logic Controllers). The example uses the terms: states and transitions; i.e., it suggests the use of the state machine concept in PLC design. The example is then realized as a state machine using StateWORKS, illustrating how the functionality can be completed and improved.

Comparing the StateWORKS solution with the approach presented in the IEC document we may formulate some reflections about the evolution taking place in the PLC world and suggest further changes encouraging a transition from intuitive to model-based design.

GRAVEL Example from IEC 61131 Document — Critical Analysis

The description of the control problem from the IEC document begins:

A control system is to be used to measure an operator-specified amount of gravel from a silo into an intermediate bin, and to convey the gravel after measurement from the bin into a truck.

287

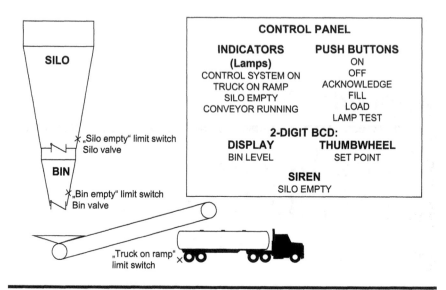

Figure E.1   Gravel system.

The required operations can be specified referring to Figure E.1.

The description suggests the designed system uses momentary-action pushbuttons (like keyboard buttons, i.e., they have one stable position). The JOG button is not shown on the console panel; we assume that it is mounted somewhere close to the conveyor and used in emergency and service cases.

The requirements are: the ON button switches on the automatic control, the OFF button stops it. If the automatic control is on, the FILL button starts filling the bin. The filling terminates automatically if the required (set by a thumbwheel) amount of gravel is in the bin. When the bin is filled and the truck is on the ramp (detected by a limit switch), the LOAD button switches on the truck loading: the conveyor motor is started and after a while when it reaches the full speed the dumping of the bin contents begins. When the bin is empty (detected by a limit switch) the conveyor runs still for a while, so that the entire load of gravel is loaded onto the truck. At any time the loading can be stopped and reinitialized if the truck leaves the ramp or if the automatic control is off.

The automatic control "on" and the presence of the truck on the ramp are each to be signaled by a lamp. The conveyor running and silo empty are to be signaled by a blinking lamp. In addition a siren is to signal that the silo is empty. The siren can be acknowledged by a button for a certain time: if the silo stays empty, the siren will be restarted.

The implementation does not mention explicitly the concept of state machines but uses the word "state" and "transition." Thus, it is relatively

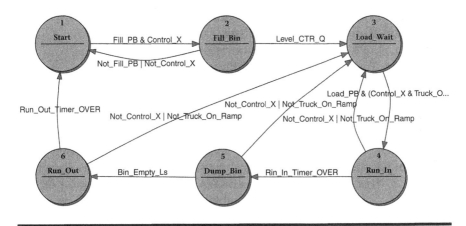

Figure E.2   IEC Gravel state machine: the ST diagram.

easy to identify the main state machine in the design. The control system is realized as SFC (Structured Function Chart) using ST (Structured Text) language elements. The SFC has two representations: a text (the actual program) and a graph (a kind of flowchart with symbols of logical elements like gates and flip-flops).

The states are declared with the keyword STEP, the initial state with INITIAL_STEP and the transition with TRANSITION. Hence the initial step looks like:

```
INITIAL_STEP START ; END_STEP

TRANSITION FROM START TO FILL_BIN

:= FILL_PB & CONTROL.X ; END_TRANSITION
```

Analyzing the SFC program we identify four state machines or we could also say three state machines and one combinational system. The first state machine (let us call it Gravel) is programmed under the title "Major operating states" and is shown in Figure E.2. This state machine changes its state as triggered by inputs but does not control anything directly (well, the declaration:

```
:= TIMER.T >= RUN_TIME:
```

effectively starts the timer and waits for the time-out to perform the transition).

To keep the correspondence between the code in the ICE document and the state transition diagram as close as possible we use the original variable names from the IEC document. The meaning of the names, which might be confusing for the reader, is as follows:

- Control_X -> Automatic Control is On
- Not_Control_X -> Automatic Control is Off
- Fill_PB -> FILL PushButton is pressed
- Not_Fill_PB -> FILL PushButton is not pressed
- Level_Ctrl_Q -> Required level in Bin is reached

Other names are more or less understandable.

If there is a "major system" we would expect there will be some "other systems" too. No, the second state machine (let us call it Control) is programmed under the title "Control state sequencing" and is shown in Figure E.3. This state machine remembers the last pushed button ON or OFF, and its state is used in the combinational system as well as in the state machine Gravel in logical conditions. The Control state machine implements the conveyor motor control and switching of the blinking lamp signaling that the silo is empty. A section drawn as two delay components assures the blink timer function, and could be considered as a state machine having only two states.

The combinational system (let us call it Actions) decodes the situation using the states of the state machines plus inputs and determines outputs. The entire control system is shown in Figure E.4.

Some comments about the implementation:

- The use of the FILL push button is not clear. If it is a 1-position pushbutton then the operator has to push the button all the time during the filling; when the operator removes his or her finger from the button, filling will be interrupted and the system returns to the state *Start*. The other solution would be to use a 2-position button.
- The control of the silo valve is shown in SFC but is missing in the program (should be opened in the state *Fill_Bin* and closed otherwise).
- The control of the bin valve is shown in SFC but missing in the program (should be opened in the state *Dump_Bin* and closed otherwise).

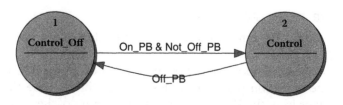

Figure E.3   IEC Control state machine: the ST diagram.

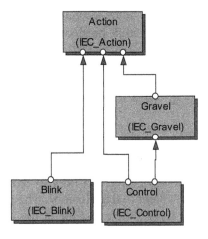

Figure E.4 The IEC Gravel control system as SFC.

■ The control of the siren signaling empty silo is shown in SFC in such a way that a hardware designer would not accept it (a flip-flop with both R and S inputs active and assuming that the input R wins). The implementation in the program is so complex we can only believe that it works.

■ The test of the conveyor lamp is missing.

■ The design has some weak points. For instance, when loading is started it cannot be interrupted but the entire content of the bin must be loaded onto the truck. The other not so nice solution is that if the run-out is interrupted and the conveyor starts again the run-out time is calculated always from the beginning.

■ The problem mentioned in the previous point may be truly unpleasant in case of limit switch malfunctions. For instance, if the BinEmpty sensor does not work there is no way to reach the initial *Start* state except by restarting the system.

■ There are no precautions for detecting and reacting to various malfunctions.

■ There is no way to switch off the siren if the silo is empty: the siren whine can be only interrupted for a while.

■ The last comment relates to the syntax of the SFC representation. I would have never managed to understand the functioning of ST language elements: Blink, Blank, and Pulse without the SFC diagram. And that was only understandable for me because of my hardware background that I still have not forgotten. I cannot save myself a (rhetorical) question why PLC programmers do not protest against that kind of user-unfriendly syntax. In general it makes an odd impression to keep at all costs a hardware diagram that should explain a totally unreadable program.

Omissions discussed above are typical for sunny-day-scenarios where we implement the sequences that should happen. The unexpected situation will be dealt with later in some way: often by code.

A State Machine as a Replacement for Markers

PLC programming has always used markers to store information about the past. Markers are equivalent to flags in code. The problem with markers is the same as with any flags: they are difficult to control if their number increases. The form of the control system in the example corresponds directly to the concept of markers: the steps (i.e., states) are in effect treated as a better way to organize markers. But effectively, the code defines a state machine.

A code in the IEC document is developed in the following way: we design a state machine that reflects a certain "sequence." Then we build a combinational circuit, which is built using the state of the state machine and inputs. The problem lies in the understanding and design of the "sequence." Such state machines are used but only rarely, under special circumstances. Readers who look at the case study "TrafficLight control" on our Web site would discover there exactly that kind of solution. For the TrafficLight control we use a state machine whose sense is to reflect the *position* and the *movement direction* of a train in the controlled zone. If we know that information, we can decide about the traffic light: it must be *on* if at least one TrafficLight state machine signals that the train moves toward the crossing (there is a separate state machine for any train in the controlled zone). The solution is simple and the design of the "sequence" is obvious. Those state machines are so-called *parser* state machines — we come back to that topic later in Conclusions where we present the state machine classification in more detail.

Most control systems are not so homogeneous considering their behavior. It is very difficult to find the actual "sequence" that assures that the outputs of the control system can be defined as a set of combinational Boolean equations. In the example, the sequence has been found but as we pointed out in the analysis it covers only the sunny-day-scenario: the full functionality requires additional effort. Therefore the normal and recommended way is to design a state machine that includes explicit output actions. This is not just a cosmetic change. The analyzed example, as presented, is a simple control problem. Simple problems can be solved using any method, intuitively rather than by developing a method-based strategy; anything will do. The true difficulties in a design arrive with increasing complexity of the control requirements.

The state machine approach requires the conviction that by knowing the state (supported by inputs when using a Mealy model) all outputs or other activities are determined. In other words, when designing a state machine we think in terms of the states: in any state we decide what to do and when to change the state. With the approach presented in the IEC document, the state machine is a supporting instrument only. If a designer of such a "marker" state machine decides for some reason to change the state machine he has to go through the entire combinational part trying to understand how the change influences the output conditions. The separation between the actions and the state machine makes the design difficult. The astonishing factor is that this difficulty is self-imposed; it is only a question of proper understanding of the role of a state machine in the system design.

The marker approach also weighs on the general approach. If we look at the program (textual SFC) it is a typical program. We change a state, we do something in the state, other actions are done in the combinational part. It may and it will work after some time: it is a question of how many hours we put into testing it. But it is very difficult to conquer a complex problem in such a way: it becomes more and more difficult to see all the dependencies.

So, we are now coming to the next problem. When designing complex control systems we need several state machines that communicate among themselves. If we need several state machines with exactly the same behavior we can do it with a "copy and paste" method; it is not a very attractive perspective. We cannot achieve a clear structure in such a coded solution.

In any case, we find this arrangement great progress in comparison with markers but we also understand the resistance that PLC programmers have against true understanding of a state machine. Unfortunately, they do not understand the vitally important distinction between using markers and using states.

GRAVEL Example as a State Machine

To support the criticisms from the previous section we have designed the control system using StateWORKS, treating the state machine as the central point of the design. To make it comparable we just took the state machine Gravel from the document as the basis. The entire system is shown in Figure E-5. In addition to the Main Gravel state machine it contains four other state machines:

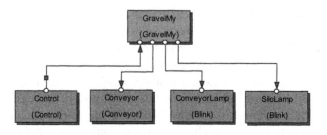

Figure E.5   Gravel control system: the SMS diagram.

- We retained the Control state machine to transform the push buttons ON and OFF into easier-to-use states: *On* and *Off*. Alternatively we could have used the state machine as a Master that sends the commands: Cmd_On and Cmd_Off to the Gravel state machine.
- There are two blinking lights required. We used for that purpose a standard state machine Blink, which we have in our library. The state machine Blink is controlled by three commands: Cmd_Off, Cmd_On, and Cmd_Blink. We use two incarnations of that state machine: SiloLamp and ConveyorLamp, which receive commands from the state machine Gravel.
- We moved the conveyor control to a separate state machine Conveyor. In this example it is not really necessary but we wanted just to support the idea of decomposing a control system into a set of specialized state machines. The Conveyor state machine does the complete motor control function: triggered by commands from the Gravel state machine or by a JOG pushbutton. If the control of the conveyor becomes more difficult or different, we just change the design of that state machine but the Gravel state machine will stay unchanged.

The heart of the system is the state machine Gravel, which controls all actions. The state transition diagram of the Gravel state machine shown in Figure E.6 is very similar to the original state machine in the IEC document.

We tried to improve the original state machine by adding things where we had found fault with the implementation in the IEC document. We made the improvements under the assumption that we can use only the existing hardware: sensors, actuators, as well as pushbuttons and lamps on the console panel. Hence, we could make changes only to the control sequences. We have changed the following items:

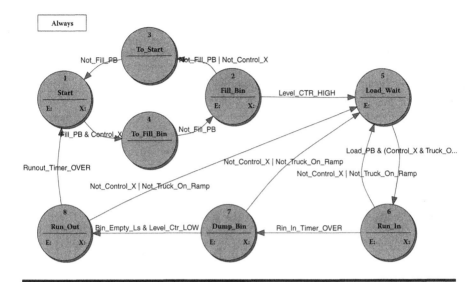

Figure E.6   My Gravel state machine: the ST diagram.

- We assumed that the FILL pushbutton is actually a momentary pushbutton exactly as the other buttons. The two additional states, *To_Start* and *To_Fill_Bin,* realize the required flip-flop effect.
- If the conveyor run-out is interrupted and later restarted, the run-out timer is not restarted but started from the already counted value. This arrangement shortens the run-out time, which otherwise is unnecessarily long. It could be improved further: we could decrease the run-out time by subtracting a value that results from the repeated run-in phase.
- The transition from the state *Dump_Bin* to *Run_Out* depends not only on the Bin_Empty limit switch but also on the supervision of the gravel level in the bin. This is an additional security, which guards against malfunction of the limit switch.
- The pushbutton LAMP TEST now tests all the lamps.
- The siren whining is limited to a few times only (a constant in the configuration, which can be changed at any time). After the third ACKNOWLEDGMENT the siren stays silent independently of the state of the silo limit switch. It "recovers" if the silo becomes "not empty" again.

The difference between the implementation in the IEC document and the state machine above is that the latter contains all actions that are under its control. Figure E.7 shows as an example the state transition table of the state *Run_In.*

Run_In	Entry action	ConveyorCmd_On ConveyorLampCmd_Blink Rin_In_Timer_ResetStart
	eXit action	Rin_In_Timer_Stop
Dump_Bin	Run_In_Timer_OVER	
Load_Wait	Not_Control_X | Not_Truck_On_Ramp	

Figure E.7 My Gravel: the ST table of the state *Run_In*.

When entering the state, the commands Cmd_On are sent to the Conveyor and ConveyorLamp state machines and the timer Run_In_Timer is restarted. When the timer elapses the state machine goes to the state *Dump_Bin*. Before the timer elapses the state machine may return to the state *Load_Wait* if the automatic control is switched off or the truck leaves the ramp. Leaving the state (independently of the destination state) the Run_In_Timer is stopped as the timer loses its meaning in other states.

There are still a few items missing or that should be corrected:

- Malfunction of limit switches should be signaled to the operator by an alarm but it would require an alarm display or additional warning lamp(s) on the console.
- The return to the initial state *Start* in case of a limit switch malfunction is not possible. Rearranging the role of the existing switches could allow it: the ON button could force the return to the state *Start* and the LOAD button could be used to stop the conveyor in states: *Run_In*, *Dump_Bin*, and *Run_Out*. Another possibility would be to expand the functionality of the ON button: pressing the button longer (e.g., 5 seconds), being treated as a break of loading, forces a return to the state *Start*. Those revisions would require approval from the client.

These further changes would make a good exercise for interested readers.

All details of the Gravel control system can be found in the StateWORKS project Gravel on our Web site. The results can be tested using SWLab and SWMon. All case studies and tools may be downloaded from www.stateworks.com.

Conclusions

Even the half-hearted use of the state machine concept in PLC programming is a useful step forward, from the intuitive way of programming by coding, to programming that is based on control system modeling. The way it is presented in the IEC example is still influenced by the "marker" way of thinking, which makes it less attractive for PLC programmers. Missing a concept of a system of state machines limits the use of state machines in PLC programming to very simple examples.

The major flaw in the IEC example is that the state machine is not considered a real solution for the application control flow. The state machine is used to store some information about the situation, which covers the *sunny-day-scenario*, while the details of system malfunctions are left for the programmer: the programmer will arrange it in some way. Problems caused by rarely occurring malfunctions are the essence of the designer's real task: the full advantages of correct state machine use are seen by solving those truly difficult control sequences with relative ease.

Several criteria can be used for classification of state machines. For instance, the Moore and Mealy model definitions come from the educational/scientific world. Taking the application criteria into consideration we would rather speak about *parser* (in automata theory it is also called *acceptor* or *recognizer*) and *control* (in automata theory it is called also *transducer*) state machines.

The *parser* state machines match strings (in general symbols); i.e., they follow a sequence of states with the purpose of detecting a certain string pattern. In practice, they do nothing else while changing the states and the concept of output actions does not make sense for them.

A specific variant of a *parser* state machine transforms changes of input signals into a set of states that can then be used by simple decoding to determine outputs. Such situations occur relatively seldom and are characterized by a specific homogenization of the control requirements, which happens only for rather simple control systems.

Both variants of *parser* state machines are special cases of state machines, finding little application in industrial control because there are not many situations where they may be used.

The *control* state machines are actual state machines used in practical applications. They consist of states that determine output actions. With *control* state machines we are able to specify behavior of any complexity, especially by using a system of state machines.

Thinking about solving a control problem by serious use of state machines allows us to use for this purpose StateWORKS Studio. With that tool we can solve the problem and test the solution. If we are sure that it works we can then code it. An automatic translation from the StateWORKS Studio specification to an SFC program is also imaginable, why not? The best solution would be, of course, to use the StateWORKS run-time system and avoid the code generation.

If the control system runs anyway on a PC (it is the case when using Beckhoff hardware) the use of StateWORKS run-time systems seems to be the natural choice. A simulation of a PLC run-time system on a PC would be in that case rather difficult to justify.

If readers study the example carefully they will see that a complete solution — within imposed limits — is presented, and not just a simplified one. It is our conviction that the complete solution to the complete problem must be produced by the design methods used, and all its behavioral features must be presented very clearly to the designer, without hiding major aspects in code or elsewhere. If StateWORKS is used, the designer is encouraged — one might almost say forced — to adopt a very healthy design methodology, thus ensuring that the software will be highly reliable in service.

Testing with SWLab

As SWLab has only eight digital input switches, they are used as push-buttons: ON, OFF; ACKNOWLEDGE, FILL, LOAD; as limit switches: BinEmpty, SiloEmpty; and as an optical sensor TruckOnRamp. The two additional inputs — LAMP_TEST and JOG — are accessible in monitors (SWLab, SWQuick, or SWTerm).

SWLab has eight digital output lamps. It is sufficient for the example and we use them for the four indicators: CONTROL_ON, TR_ON_RAMP, SILO_EMPTY, CONV_RUNNING; and for the four outputs: SiloSiren, Silo-Valve, BinValve, ConveyorMotor.

From the four analog inputs that SWLab has we need only one: GravelLevel.

The rest of the objects used: timers, gravel SetPoint as a parameter, switchpoint GravelSwip to detect the gravel level in the bin, as well as state machines and their commands are accessible in StateWORKS monitors.

Recommended Reading

1. International Standard IEC 61131-3, 2nd ed. 2003-0.

Appendix F

Traffic Light Control — Design of the Hardware Solution

Discussing the hardware solution for a Traffic light control (see Figure 5.2) we have skipped the creation of Boolean equations for the D flip-flops. The Karnaugh tables for the flip-flops inputs, shown in Table F.1 and Table F.2, are based on the state coding as set in Table 5.1 (see *Design example — traffic light control* in Chapter 5).

Table F.1   Karnaugh Tables of D Flip-Flops

$Q_2Q_1Q_0$	XM 00	01	11	10
000	0	-	-	0
001	0	-	-	0
011	1	0	-	-
010	0	0	-	-
110	-	-	-	-
111	-	-	-	-
101	0	-	-	1
100	1	-	-	1

D_2

$Q_2Q_1Q_0$	XM 00	01	11	10
000	0	-	-	0
001	1	-	-	0
011	0	1	-	-
010	1	1	-	-
110	-	-	-	-
111	-	-	-	-
101	0	-	-	0
100	0	-	-	0

D_1

$Q_2Q_1Q_0$	XM 00	01	11	10
000	0	-	-	1
001	0	-	-	1
011	0	1	-	-
010	0	1	-	-
110	-	-	-	-
111	-	-	-	-
101	0	-	-	1
100	0	-	-	1

D_0

Table F.2   Karnaugh Table of the Output Signal Y

Q_2	Q_1Q_0 00	01	11	10
0	0	1	1	1
1	0	0	-	-

Y

Appendix G

Coding Finite State Machine — Vending Machine Counter Example

```
// VendingMachine.cpp
//
// The program demonstrates the use of a coded state
// transition table for implementation of a state machine.
// When running the Console asks for Input.
// Values: 5 and 10 cause the state machine change the
// state according to the state transition table in the
// Figure 4.3 (chapter 4), other values are ignored.
// In state Stop (value 25 is reached) any value different
// than 5 or 10 changes the state to Start and the exercise
// can be repeated.

#include "stdafx.h"
#include "stdio.h"

using namespace std;
```

```
enum eState
{
   Start, Five, Ten, Fifteen, Twenty, Stop
};

char* sState[Stop+1] =
{"Start", "Five", "Ten", "Fifteen", "Twenty", "Stop"};

enum eInput
{
   coin5, coin10, coin
};

eState transition[Stop+1][coin+1] =
{//           "coin5"     "coin10"     "coin"
/*Start*/     {Five,      Ten,         Start},
/*Five*/      {Ten,       Fifteen,     Five},
/*Ten*/       {Fifteen,   Twenty,      Ten},
/*Fifteen*/   {Twenty,    Stop,        Fifteen},
/*Twenty*/    {Stop,      Twenty,      Twenty},
/*Stop*/      {Stop,      Stop,        Start}
};

int _tmain(int argc, _TCHAR* argv[])
{
   eState iState = Start;
   eInput iInput;
   int iTemp;

cout << endl << "Input = ";
   cin >> iTemp;
   while (iTemp != 'q')
    {
      switch (iTemp)
      {
        case 5:
           iInput = coin5;
           break;
        case 10:
           iInput = coin10;
           break;
```

```
        default:
            iInput = coin;
            break;
    }
    iState = transition[iState][iInput];
    cout << sState[iState] << endl;
    cout << endl << "Input = ";
    cin >> iTemp;
    }
    return 0;
}
```

Appendix H

IOD File of the StandardUnit

H ...\Standard\Conf\StandardUnit.iod

B #	Name	Object List	Type	Description
1	Par_DllName		11	must-not-be-changed
2	Al_ReadDiError		4	must-not-be-changed
3	Al_ReadXdaError		4	must-not-be-changed
4	Al_ReadNiError		4	must-not-be-changed
5	Al_ReadDatError		4	must-not-be-changed
6	Al_WriteDoError		4	must-not-be-changed
7	Al_WriteCmdError		4	must-not-be-changed
8	Al_WriteNoError		4	must-not-be-changed
9	Al_WriteTabError		4	must-not-be-changed
10	Cmd 2			
11	Di_0 5			
12	Di_1 5			
13	Di_2 5			
14	Di_3 5			
15	Di_4 5			
16	Di_5 5			

```
17     Di_6 5
18     Di_7 5
19     Do_0 6
20     Do_1 6
21     Do_2 6
22     Do_3 6
23     Do_4 6
24     Do_5 6
25     Do_6 6
26     Do_7 6
27     Ni_0 8
28     Ni_1 8
29     Ni_2 8
30     Ni_3 8
31     No_0 7
32     No_1 7
33     No_2 7
34     No_3 7
35     Dat_1 15
36     Dat_2 15
37     Xda_0 10
38     Xda_1 10
39     Tab 19
40     Par_PollingTime                11    must-not-be-changed
```

C #	Name	Cmd List	Value
1	STDC_Cmd1		1
2	STDC_Cmd2		2
3	STDC_Cmd3		3
4	STDC_Cmd4		4

Appendix I

StateWORKS Projects

A Project

A StateWORKS project covers two specifications:

- state machine (VFSM) and UNIT types
- system

Both specifications are separate and may be made in any sequence. The VFSM and UNIT type are abstract specifications. The state machine specification defines a new object of a type VFSM. The UNIT specification defines a new object of a type UNIT.

The system specification is a specification of the RTDB using all available objects in the project: the predefined objects and the project specific VFSM and UNIT objects.

Specifications produce several files. All those files are stored in a common project folder with a default name *Conf*; the user may define any folder path.

The projects may be organized in the following directory structure:

```
Projects
    ProjectName1 (contains *.prj file)
        Conf (contains *.swd file and all *.iod, *.str, *.h files)
        Xml (contains *.xml, vfsmml.xsl, vfsmml.dtd files)
            Graphics (contains *.jpg or *.wmf files)
    ...
```

ProjectNameN
   Conf
   Xml
      Graphics
   VFSM (contains all *.fsm)
   UNIT (contains all *.unt)

We recommend this kind of arrangement as it makes easier the re-use of
VFSM and UNIT types in several projects.

Specification of State Machines

Specification of a state machine begins with definitions of the virtual
environment, which comprises three sets of names: Input, Output, and
State. The names directories may be changed, reduced, and expanded at
any time on demand. The Input and Output names are created on
predefined object types plus the states and commands of slaves (VFSM)
included in the project.

The specified state machine is a new object type and can be used for
a system (RTDB) specification exactly as for predefined object types.

The results of the specification are in three files:

- **IOD**: contains list of Object Names (**B** section), Input Names
  (**I** section), Output Names (**O** section), States Names (**S** section),
  and CMD Names (**C** section)
- **STR**: contains the behavior specification of the state machine
- **H**: is a C/C++ h-file containing the same information as the IOD
  file but as C enumerations

The IOD and STR files are read by the RTDB-based runtime system. The
H file is used by programming of I/O handlers or output functions. All
files are text files and can be read by any text editor. Their content must
not be changed by hand. All files are unique for any VFSM object of that
type used in the project; in other words, they exist only once for all VFSM
instances of that type (the behavior of a state machine of a given type
stays the same — only the objects used by the state machine instances
are different).

Note that a state machine specification is completely independent from
a system specification; it is actually a definition of behavior in a virtual
environment.

Specification of UNITs

Specification of a UNIT means definition of an object list. The list may contain any number of predefined (RTDB) object types.

The specified UNIT is a new object type and can be used by a system (RTDB) specification exactly as for other object types.

The results of the specification are two files:

- **IOD**: contains list of Object Names (**B** section) and CMD Names (**C** section)
- **H**: a C/C++ h-file containing the same information as the IOD file as C enumerations

The IOD files are read by the RTDB-based runtime system. The H file is used by programming of I/O handlers or output functions. All files are text files and can be read by any text editor. Their content must not be changed by hand. All files are unique for any UNIT object of that type used in the project; in other words, they exist only once for all UNIT instances of that type.

Note that a UNIT specification is completely independent of a system specification; it is actually a definition of object lists that might be used to organize the access to I/O handlers and output functions.

System Specification

The system specification is a specification of objects used by the project. It should contain at least all objects needed by state machines, I/O handlers, and output functions. In addition, we may add any objects we need for storing application data. Several of those additional objects are indirectly linked with the state machines; for instance, if the SWIP object used in a state machine requires limits defined by parameters we specify PAR objects for that purpose (which do not belong to the state machine).

The specification covers definitions of object properties, which for several objects means a definition of links between objects:

- Master and slave state machines (VFSM and CMD)
- UNIT lists
- Control values (states) of counted objects for event counters (ECNT)
- Parameters (PAR, DAT) for: timers (TI) time-out, switchpoints (SWIP) limits and input, numerical outputs (NO) output, tables (TAB) list, strings (STR) input, and regular expression source.

Only if all those properties are defined do we get the entire picture of the system, among others, the System of State Machines (**SMS**) diagram.

The result of the system specification is a configuration file **SWD**.

Note that although a system specification is not completely independent of the state machine and UNIT specifications it may be done at any time. For an application for which the I/O system is well defined we may specify all input and output objects that are expected to be needed in the application. Such an RTDB may be used as the application in a first instance, just to test the hardware before or in parallel to creation of the behavior (state machine) specifications, which will be added later.

Documentation

There are several ways to study a StateWORKS project. Probably the easiest way is to load the project into the StateWORKS Studio where we can access and display all details of specifications: state machines, UNITs, and system.

The specification results — SMS diagram, ST diagrams, ST tables — can be saved as *.jpg or *.wmf files to be included in documents. Similarly, all specification files are text files and can be displayed or included in various documents.

XML files represent also a specific form of documentation. There are XML files for all state machines as well as a general file that contains the entire knowledge about the project: all state machines, UNITs, and system configuration. The XML files are generated automatically when building the entire system of state machines in a default subfolder of the project directory *Xml*. The automatically attached files vfsmml.xsl and vfsmml.dtd allow their content to be displayed in any Web browser. These files can be taken as guidance and a new XSL file edited, to create special versions of documentation, or for converting a StateWORKS project for entry to some other application.

Testing with SWLab and Monitors

The developed system can be tested using the simulator SWLab. SWLab is an application built on the RTDB. Therefore state machine behavior is modeled in SWLab exactly as in the ultimate application. The main advantage of using SWLab is its user interface, which simulates the basic I/Os: 8 digital inputs (DI), 8 digital outputs (DO), 4 numerical inputs (NI), and 4 numerical outputs (NO). An application that wants to use the simulated I/Os must be built with standard UNITs of a type: DI8, DO8,

NI4, and NO4. Of course, the limited number of I/Os limits the use of simulated I/Os to rather simple applications; e.g., all examples discussed in the book can be tested with SWLab. Therefore the examples are built with the standard UNITs for SWLab.

We may use SWLab to test applications that require any number of I/Os of any sort. In such a case we just ignore the SWLab user interface with the simulated I/Os and change the I/Os directly in the RTDB using Monitors; StateWORKS development system has three: SWMon, SWTerm, and SWQuick.

The ultimate test will be done using an application that is built (similarly to SWLab) on the RTDB but contains the true I/O handlers that realize the hardware interface.

In any case, in testing, the central role is played by Monitors. These monitors allow us to have access to all objects used in the systems: state machines, commands, alarms, timers, parameters, counters, switchpoints, output functions, etc. To list the most important features we may:

- Watch the cooperation among state machines using the command/ state interface.
- Change the values of several objects trying to identify all imaginable control problems.
- Watch the objects in different relations: all, state machine (VFSM), or UNIT.
- Start several monitors to watch the system from different perspectives at the same time.
- Investigate the state machine problems using run or step mode.
- Debug errors using trace and command file.

Monitors also play a very important role while supervising the operating control system — they provide information which we might get using a code debugger in a programmed application.

Documentation of Examples

In the following appendices we provide documentation of all projects that contain the state machines discussed in the book. The characteristic of a project above shows that the documentation of StateWORKS specification is created automatically but it is voluminous. To provide all created files and diagrams in a printed form would greatly expand the appendices and the book. Therefore we limit the printed matter to the most important diagrams that explain the solution, supported by a description containing some essential information. The size of the explanation is adjusted to the difficulty of the problem. We have concentrated especially on description of testing: test routines are not part of the automatically created documentation, but they are important when analyzing a project, especially if done by another person. Normally, various command files and log files would be generated during testing, and preserved with the other documentation.

The complete projects are available on the Internet (www.stateworks.com). The StateWORKS Studio software provided on the Internet allows analysis of the examples to the last details, including making and studying changes if the reader wishes to do that. (You can download the projects and software from www.stateworks.com. As a buyer of the book you are entitled to free registration of the software, which can then be used without a time limit. To register, use the serial number printed on the inside of the book's back cover.)

Appendix J

Vending Machine Counter Project

The behavior of the vending machine counter is described by the state transition diagram in Figure J.1. It covers only the sunny-day-scenario: that is, it reacts to coins of 5 and 10 values. To achieve a functioning system that can be tested using SWLab we decided to use as a coin-input the XDA object. The object will have three values: 5, 10, and 0, which are used for creating the Input names: 5, 10, and Done. In such a way we are able to simulate the coin: after a valid coin value 5 or 10 the value 0 terminates that coin signaling that the next coin is awaited. The Done value in the state *Stop* causes a return to the state *Start,* which allows a repetition of the test.

In addition to the starting state *Start* and terminating state *Stop* the state machine has four state pairs like *Ten_Busy* and *Ten.* The state *Ten_Busy* means that the actual coin increased the counter value to 10; the state *Ten* means that the last coin has been "consumed" and the state machine is waiting for the next coin. The ST tables for those states are shown in Figure J.2 and Figure J.3. Other state pairs have a similar content. The entire project can be executed with SWLab. The values for the coin-input could be set in any StateWORKS Monitor.

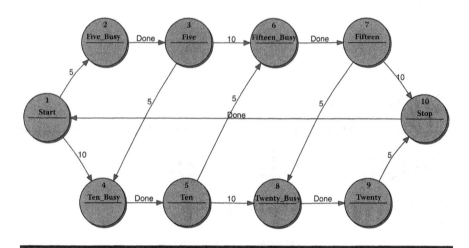

Figure J.1   VendingMachineCounter: the ST diagram.

Ten_Busy	Entry action	
	eXit action	
Ten	Done	

Figure J.2   VendingMachineCounter: the ST table of the state *Ten_Busy*.

Ten	Entry action	
	eXit action	
Fifteen_Busy	5	
Twenty_Busy	10	

Figure J.3   VendingMachineCounter: the ST table of the state *Ten*.

Appendix K

Pedestrian Traffic Light Project

The System

The requirements are formulated in *The requirements* section in Chapter 8 (see also Figure 8.6). The control system Pedestrian shown in Figure K.1 contains:

- State machine PedestrianLight of a type Pedestrian
- UNITs: DI8:01 of a type DI8 and DO8:01 of a type DO8

The State Machine of Type Pedestrian

The state transition diagram is shown in Figure K.2.

The table *Always* and the state *Init* are irrelevant and do not contain any expressions. The state machine deviates slightly from the model discussed in *Example — Pedestrian traffic lights* in Chapter 8:

- The additional state *MyInit* has been introduced to delay the access to Hardware required by the I/O handler. If the system is tested with SWLab, that state is actually unnecessary.
- Each Do has separate *set* and *reset* output signals according to flip-flop features of RTDB DO objects.

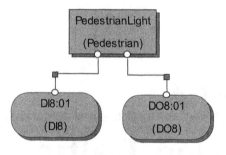

Figure K.1 The Pedestrian control system.

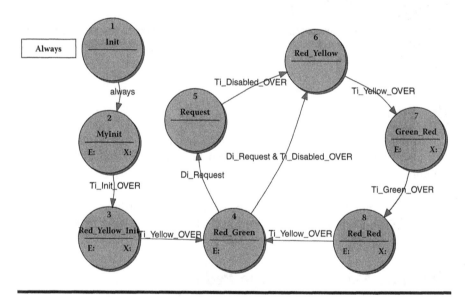

Figure K.2 Pedestrian traffic light: the ST diagram.

Testing with SWLab

As the requirements for pedestrian crossing lights are well defined in *Example — Pedestrian traffic lights* in Chapter 8, testing should not present any problem. Starting SWLab we see the button Di_Request and the crossing lights: Do_CRed, Do_CYellow, Do_CGreen for the car traffic and Do_PRed, Do_PGreen for the pedestrian. For the first 20 seconds after start the Di_Request is disabled but we may activate it — this will be stored and when the time elapses the crossing lamps work as desired.

Actually we do not need a monitor but we may start SWMon to watch the behavior of all objects, especially those that cannot be seen on the

SWLab user interface, such as states of the state machine Pedestrian and four timers. For completeness we describe the role of the timers:

- Ti_Disabled: determines the dead time after pedestrian green light (default 20 seconds).
- Ti_Green: determines the green light for pedestrian (default 10 seconds).
- Ti_Init: determines the delay after start-up (default 0.1 second).
- Ti_Yellow: determines the yellow light for car traffic (default 3 seconds).

Appendix L

Pumps Supervision Project

The System

The control requirements have been formulated in the *Example - Pumps supervision system* section in Chapter 9. Several parts of the project have been discussed in the book. The complete system shown in Figure L.1 contains:

- A state machine of a type Main: Main
- Two state machines of a type Pressure: Pressure1 and Pressure2
- A state machine of a type Device: Device1
- UNITs for I/O simulation in SWLab: DO8:01 of a type DO8, DI8:01 of a type DI8, NO4:01 of a type NO4, NI4:01 of a type NI4
- UNITs for output functions of a type OfuLimit: OfuLimit:01 and OfuLimit:02

The state machine types Main and Pressure have been discussed exhaustively in Part II (see *Example — Pressure supervision* in Chapter 8 and *Example — Pumps supervision system* in Chapter 9). Hence, we show below for completeness only the SMS diagrams of Main and Pressure and limit our description to the state machine Device.

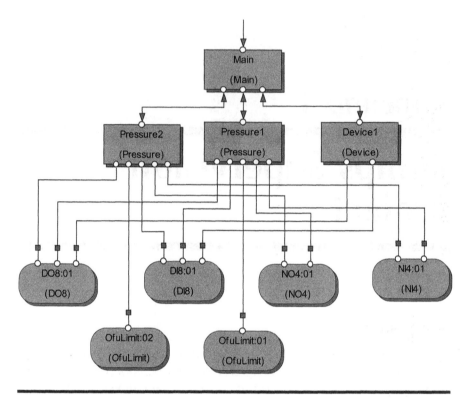

Figure L.1   The Pumps system: the SMS diagram.

The Main State Machine

The state transition diagram is shown in Figure L.2.

The State Machine of Type Pressure

The state transition diagram is shown in Figure L.3. The system uses two of them.

The State Machine of Type Device

The state transition diagram is shown in Figure L.4.

The state machine Device is a simplified but typical example of switching something on and off with a feedback from the controlled device.

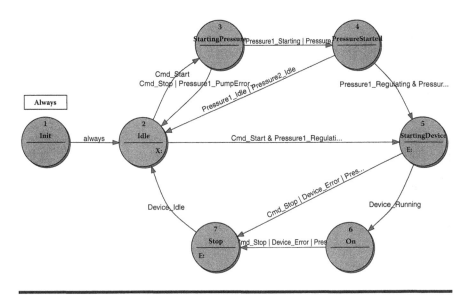

Figure L.2   Pumps: the Main ST diagram.

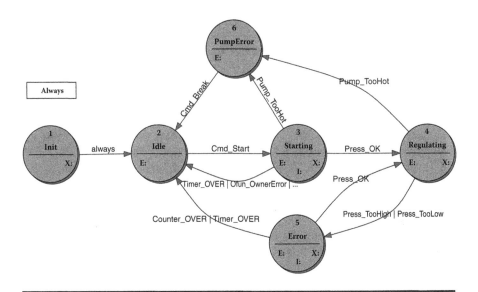

Figure L.3   Pumps: the Pressure ST diagram.

The state machine has one *busy* state: *Starting* and 3 *done* states: *Idle*, *Running*, and *Error*. The *Init* state does not play any role: the system starts in *Init* state and goes immediately to the state *Idle*, where it waits

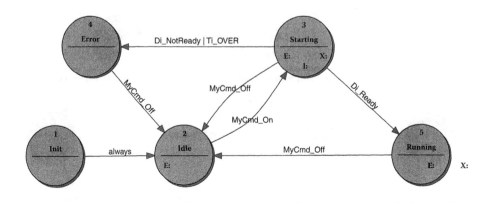

Figure L.4 Pumps: the Device ST diagram.

for the command Cmd_On. Receiving the command Cmd_On the state machine goes to the state *Starting*:

On entering the state the output Do is set to On and the command is cleared. In addition, the Ti timer is started to guard the state machine against deadlock — the state machine waits here for the feedback Di from the device. If the feedback is Ready the state machine goes to the state *Running*; otherwise (NotReady or Ti_OVER) it goes to the state *Error*. The process may be interrupted by the command Cmd_Off, which returns the state machine to the state *Idle*. Ti_OVER and NotReady feedback generate corresponding alarms as Input Actions.

State *Running* is a final required state: the device is running. A command Cmd_Off switches off the device (in the state *Idle*). The feedback from the device is not checked, which may cause problems. To be completely on the safe side we should introduce a busy state *Off_Busy*, which controls the situation in a similar way to the state *Starting* by switching on the device (Do Off, timer and feedback). The state *Idle* in the current solution may signal the master wrong information if the device has not switched off for some reason.

State *Error*: we do nothing there (note that the alarm has been sent already on leaving the state *Starting*; otherwise we would need two error states). The state is used to signal the master about problems with Device: the information does not contain details (reasons) about malfunctions — those were handled (Alarms) by the state machine Device.

Testing with SWLab

In SWLab we have access to all inputs:

- Pump temperature sensors represented by switches: Pressure1:Di: PumpTooHot and Pressure2:Di:PumpTooHot
- Device ready sensor represented by two switches: Device:Di:Ready and Device:Di:NotReady
- Actual pressure values represented by analog potentiometers: Pressure1:Ni:ActualPressure and Pressure2:Ni:ActualPressure

and we may watch all outputs:

- Pressure ok signals represented by LEDs: Pressure1:Do:PressureOk and Pressure2:Do:PressureOk
- Device *On* state represented by: Device:On
- Required pressure values represented by gauges: Pressure1:No: RequiredPressure and Pressure1:No:RequiredPressure

If we want to investigate all details of the control, monitors are very helpful. Testing of Pumps systems is quite complex and demonstrates well control problems that may occur in industrial practice.

By testing we have to take into consideration all dependencies that originate from links in the project and other decisions taken by design particularly:

- The switchpoints: Pressure1:Swip:ActualPressure_Supervison and Pressure2:Swip:ActualPressure_Supervison supervise, respectively, Pressure1:Ni:ActualPressure and Pressure2:Ni:ActualPressure.
- Default limits for switchpoints are 0 values as they are calculated by calling output functions.
- The pressure values Pressure1:No:RequiredPressure and Pressure2: No:RequiredPressure are determined by parameters correspondingly: Pressure1:Par:RequirePressure and Pressure2:Par:RequirePressure.
- The event counters Pressure1:Ecnt:Counter and Pressure2:Ecnt: Counter count the state *Error* of correspondingly state machines: Pressure1 and Pressure2.

All this information has been defined in the project and is available in the configuration file Pumps.swd.

Other sources of information regarding the links between real values and names in virtual environment used by state machines specification are included in IOD files for state machine types: Main, Pressure, and Device.

Of course, starting the project Pumps in StateWORKS Studio is the easiest way to display, investigate, and change those values and links.

Appendix M

Output Function CalcLimits()

```cpp
// FILE NAME : ofulimit.cpp
//
// CREATION   : 23-Oct-2003
//
// EDIT HISTORY
//
// SUMMARY
// Impementions of the user written output funtions.
//
#include "OfuLimit.h"

#include "vsofun.h"
#include "vspar.h"
#include "vsswip.h"

const int iDEV_DEFAULT = 1;
const float fDEV_DEFAULT = 0.05;

const int iINCORRECT_OWNER = 0;
const int iRESULT_OK = 1;
const int iWRONG_PARAMETERS = 2;
```

```
//-------------------------------------------------
    int CalcLimit (CItem* pOwner, int nVO)
/*
```

The CalcLimit() function is used by an OFun object
  to calculate limits for SWIP object which supervises
  a value.

Specifying the SWIP object the LimitLow and LimitHigh
  values should be left as default (By_Value = 0).
  The CalcLimit() function called by the OFun object
  will set the required value.

The objects required by the CalcLimit() function are
  specified in the Unit OfuLimit.unt:

- Swip: the SWIP object
- Par_Value: the reference value for limits calculation. It maybe a float or int type.
- Par_Deviation: the actual limit. It must be a
  string in the form: x, x%, where x is an integer
  or a float number.

The function awaits the parameter value nVO=1.

The limits are calculated as:

```
    LowLimit = Par_Value - x
    HighLimit = Par_Value + x
```
or
```
    LowLimit = Par_Value - 0.01*x*ParValue
    HighLimit = Par_Value + 0.01*x*ParValue
```

The function returns:

0: if the Function Owner (state machine) is not
correct.

1: if the calculation results are correct.

2: if the function parameter is not correct.
```
*/
//-------------------------------------------------
{
    if (!pOwner) return iINCORRECT_OWNER;

if (nVO == 1)
{
```

```
  C_PAR* pPar = (C_PAR*)(pOwner->
   GetAssItem(LIMB_Par_Value));
  C_PAR* pParDev = (C_PAR*)(pOwner->
   GetAssItem(LIMB_Par_Deviation));
  C_SWIP* pSwip  = (C_SWIP*)(pOwner->
   GetAssItem(LIMB_Swip));

  if ((pPar == NULL)||
     (pPar->GetType() != IT_PAR) ||
     (pSwip == NULL) ||
     (pSwip->GetType() != IT_SWIP)
    )
    return iINCORRECT_OWNER;

  float fDev = fDEV_DEFAULT;
  float fVal = pPar->fGetData();
  if ( pParDev != NULL )
  {
   string sDev = pParDev->stGetData();
   int iPos = sDev.find('%');
   if ( iPos < 0 )
   { // Par_Deviation is an integer
     fDev = atof(sDev.c_str());
   }
   else
   { // Par_Deviation is a string expressing the
     // limits as % of Par_Value
     sDev = sDev.substr(0,iPos);
     fDev = 0.01*atof(sDev.c_str())*fVal;
   }
   float fLimH = fVal + fDev;
   float fLimL = fVal - fDev;
   pSwip->SetLimits(fLimL, fLimH);
  }
  return iRESULT_OK;
 }

 return iWRONG_PARAMETERS;
} // End of CalcLimit
```

Appendix N

Traffic Light Project

The System

The control requirements have been formulated in *Example — Traffic light control* in Chapter 9. Several parts of the project have been discussed in the book. The complete system for two rails shown in Figure N.1 contains:

- Four state machines of a type Light: Light:0:Left, Light:0:Right, Light:1:Left, and Light:1:Right
- A state machine of a type Flash: Flash
- A state machine of a type TrafficLight: TrafficLight
- UNITs for I/O simulation in SWLab: DO8:Unit3 of a type DO8, DI8:Unit1 of a type DI8.

The Flash State Machine

The state transition diagram is shown in Figure N.2.
   The objects are represented by:

- Cmd: Flash:Cmd of a type CMD
- Light: Flash:Do:Light of a type DO
- Timer1: Flash:Ti:Timer1 of a type TI
- Timer2: Flash:Ti:Timer2 of a type TI

The time-outs are constant values defined as timer properties.

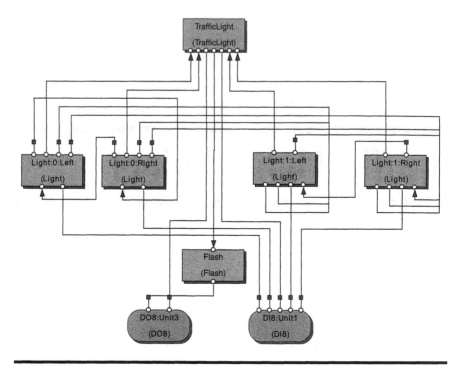

Figure N.1 The TrafficLight system.

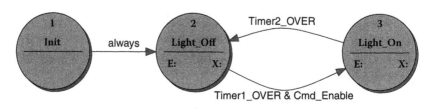

Figure N.2 TrafficLight: the Flash ST diagram.

The TrafficLight State Machine

This is a combinational control system whose behavior is described by the table *Always* in Figure N.3.

The objects are represented by:

■ Di_Light: Traffic:Di:Light of a type DI
■ Left_1: Light:0:Left of a type VFSM
■ Left_2: Light:1:Left of a type VFSM
■ Right_1: Light:0:Right of a type VFSM

- Right_2: Light:1:Right of a type VFSM
- Do_Light: Traffic:Do:Light of a type DO
- Flash: Flash:Cmd of a type CMD

The TRAFFICLIGHT state machine is in fact a combinatorial systems "decoding" states of LIGHT state machines to produce signals which switch on/off the red and yellow flashing light.

The used .._LightOn, .._LightOff, .._FlashEnable and .._FlashDisable names represent complex conditions - OR expressions of LIGHT states (see Input Name Dictionary). In these complex expressions the set of states: Coming, Approaching, ApprPresent, Present, AllPresent LeavePresent is represented only by a state Coming. This simplification can be done as this sequence of states is always entered via the state Coming (see state transition diagram).

Always	Di_Light_On | Left_1_LightOn | Right_1_LightOn | Left_2_LightOn | Right_2_LightOn	Do_Light_On
Always	Di_Light_Off & Left_1_LightOff & Right_1_LightOff & Left_2_LightOff & Right_2_LightOff	Do_Light_Off
Always	Left_1_FlashEnable & Right_1_FlashEnable & Left_2_FlashEnable & Right_2_FlashEnable	Flash_Cmd_Enable
Always	Left_1_FlashDisable | Right_1_FlashDisable | Left_2_FlashDisable | Right_2_FlashDisable	Flash_Cmd_Disable

Figure N.3 TrafficLight: the table *Always*.

The Light State Machine

The state transition diagram is shown in Figure N.4 (a copy of Figure 9.23 in section *Light* in Chapter 9).

There are four state machines of a type Light. For the state machine Light:0:Left the objects are represented by:

- Sensor_1: Light:0:Di:L of a type DI
- Sensor_2: Light:0:Di:R of a type DI
- Sensor_M: Light:0:Di:M of a type DI

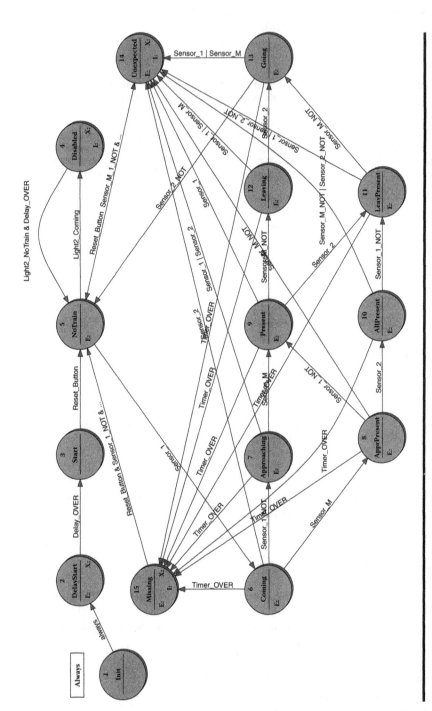

Figure N.4   TrafficLight: the Light ST diagram.

- Reset: Light:Di:Reset of a type DI (common for all Light state machines)
- Timer: Light:0:Left:Ti:Timer of a type TI
- Delay: Light:0:Left:Ti:Delay of a type TI
- Al_ActiveSensor: Light:0:Al:ActiveSensor of a type AL (common for both Light:0:Left and Light:1:Right state machines)
- Al_MissingSensor: Light:0:Al:MissingSensor of a type AL (common for both Light:0:Left and Light:1:Right state machines)
- Al_UnexpectedSensor: Light:0:Al:UnexpectedSensor of a type AL (common for both Light:0:Left and Light:1:Right state machines)
- Light2: Light0:Right of a type VFSM (the state machine for the other direction)

The objects for the other three state machines can be defined similarly and are found in the configuration file.

In addition there is one object used indirectly by all Light state machines:

- Light:Par:Timeout of a type PAR (defines time-outs for all Light Timers)

Note the naming convention: the Sensor1 and Sensor2 are numbered because for one direction they are a L(eft) sensor and for the other direction a R(ight) sensor. Therefore, for the state machine specification we use a neutral number as the specification is valid for both directions. The true sensors used in the state machine instances get then the meaning corresponding to the actual direction.

Testing with SWLab

In SWLab we have access to all input/output objects that are in that case digital inputs and outputs only. Hence, we can test the system without a monitor, especially when we are very familiar with requirements. Of course, we need a monitor if we want to watch all participating objects: alarms, timer, and the state machines.

Appendix O

DI_DO Project

The Project

The project shown in Figure O.1 contains:

- A state machine of a type Tank: Tank:01
- A state machine of a type Test_DI_DO: Test_DI_DO:01
- UNITs for I/O simulation in SWLab: DO8:01 of a type DO8, DI8:01 of a type DI8

The State Machine Test_DI_DO

This is a combinational control system whose behavior is described by the table *Always* in Figure O.2. To test the system we have to start SWLab and a monitor, e.g., SWMon. The objects are represented by:

- Button: a switch Button of a type DI
- LED: a lamp LED of a type DO

The role of the UNKNOWN value can be observed well after start-up of the SWLab when neither of the conditions is due because the application has not yet received any signal change and the VI is empty (apart from always containing the entry always). Note that we may in SWLab simulate the value UNKNOWN of the digital input: right mouse double-click or start-up options (see SWLab Help for details).

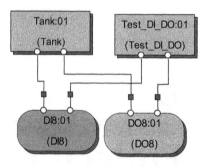

Figure O.1   The DI_DO system.

Always	Button_LOW	LED_On
Always	Button_HIGH	LED_Off

Figure O.2   Test_DI-DO example: the table *Always*.

Depending on the start options after start-up, VI of DI_DO_Test will contain the value {1} (1 means always) or {1, 3} and the DI Button should display the value UNKNOWN or 0.

Activating Button introduces a corresponding entry into VI. Of course, after the first activation VI will always contain either the entry {1, 3} (3 = Di_LOW) or {1, 2} (2 = Di_HIGH) because each change of the DI input will cause a corresponding change in VI. The values of Input names can be found in the DI_DO_D.IOD file.

The State Machine Tank

This is a combinational control system whose behavior is described by the table *Always* in Figure O.3. The requirements have been defined in *Setting and clearing the Boolean output are two different actions* in Chapter 11.

Always	DiWaterHigh_HIGH	DoValve_Close
Always	DiWaterHigh_LOW & DiWaterLow_LOW	DoValve_Open

Figure O.3   Tank state machine: the table *Always*.

The objects are represented by:

- DiWaterHigh: a switch DiWaterHigh of a type DI
- DiWaterLow: a switch DiWaterLow of a type DI
- DoValve: a lamp Do_Valve of a type DO

After starting the inputs of the state machine Tank, DiWaterHigh and DiWaterLow are either UNKNOWN or 0. Their first change in SWLab sets their value to 1. Due to the flip-flop feature of the output object DO the Do_Valve opens (1) if both inputs are LOW (0). Do_Valve is closed (0) if both inputs are HIGH (1) (effectively only DiWaterHigh plays a role — DiWaterLow must be HIGH in such a situation).

The reader may decide what to do if the inputs are UNKNOWN or what should be done if the "impossible" situation occurs: DiWaterHigh=HIGH and DiWaterLow=LOW.

Appendix P

Other_Inputs Project

The Project

The project shown in Figure P.1 contains:

- A state machine of a type Test_DAT: Test_DAT
- A state machine of a type Test_SWIP: Test_SWIP
- A state machine of a type Test_STR: Test_STR
- UNITs: DI8 and DO8 to simulate digital inputs and display digital outputs in SWLab.

The State Machine Test_DAT

This is a combinational control system whose behavior is described by the table *Always* in Figure P.2. To test the system we have to start SWLab and SWMon. The objects are represented by:

- Dat: Dat:01 of a type DAT
- Di: a switch Di:ClearDo of a type DI
- LED: a lamp LED:DAT_CHANGED of a type DO

After start the DAT object (Dat:01) is in a state INIT. Any change of DAT value is signaled by its state CHANGED. If the DAT state is used by another object (like in this example as an Input in Swip:01), that state is consumed and the DAT object goes immediately to the state DEF. The CHANGED

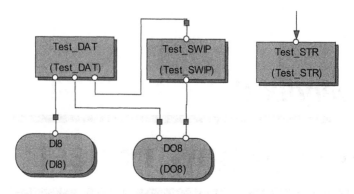

Figure P.1 The Other_Inputs system.

Always	Dat_CHANGED	LED_Dat_Changed
Always	Di_HIGH	LED_Off

Figure P.2 Test_DAT state machine: the table *Always*.

state is used to switch on the LED:DAT_CHANGED. Thus, the lighting of
LED:DAT_CHANGED proves that the condition CHANGED has been present
for a while even if we have seen nothing in SWMon. We can see the state
changes by switching on Trace or disconnecting for a while the Dat:01
object from the Swip:01. To repeat the exercise we can switch the
LED:DAT_CHANGED using the DI input Di:ClearDo.

The State Machine Test_SWIP

This is a state machine whose behavior is described by the ST diagram
in Figure P.3. The state *MyInit* has only one function: to enable the
switchpoint (see Figure P.4). After start-up the state machine always goes
from the state *Init* to the state *MyInit*.

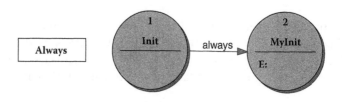

Figure P.3 Test_SWIP: the ST diagram.

MyInit	Entry action	Swip_On
	eXit action	

Figure P.4   Test_SWIP: the ST table of the state MyInit.

Always	Swip_IN	LED_In
Always	Swip_HIGH | Swip_LOW	LED_High_or_Low

Figure P.5   Test_SWIP: the table *Always*.

Control of the LED, which is the main goal of the test, is specified in the table *Always* (see Figure P.5).

The objects are represented by:

■ Dat: Dat:01 of a type DAT
■ Swip: Swip:01 of a type SWIP
■ LED: a lamp LED:SWIP_IN of a type DO

To test the system we have to start SWLab and SWMon. After start the SWIP object (Swip:01) is in the state LOW as its limits are Low Limit = 4, Limit High = 7, and the DAT object data (Dat:01) supervised by the switchpoint equals 0. Changing the DAT data we may watch the changes of SWIP state; the value IN is signaled by the LED:SWIP_IN.

The State Machine Test_STR

This is a state machine whose behavior is described by the ST diagram in Figure P.6. The entire control is specified in the state *MyInit* (Figure P.7).

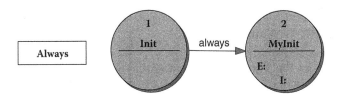

Figure P.6   Test_STR: the ST diagram.

MyInit	Entry action	Str_On
	eXit action	
	Cmd_Set	Str_Set MyCmd_Clear

Figure P.7   Test_STR: the ST table of the state MyInit.

The objects are represented by:

- Cmd: Cmd:TestSTR of a type CMD
- Str: Str:Command of a type STR

Indirectly the state machine also requires:

- Input string for the Str:Command object: Par:Input of a type PAR
- Destination objects for the Str:Command: Par:Command and Par:DeviceNumber, both of a type PAR

To test this we do not use the elements of the SWLab user interface — the objects can be accessed only from a Monitor. The test should show the functioning of the STR object. After start-up the state machine goes always from the state *Init* to the state *MyInit*. The control task is simple: the command Set should enable the STR object for another matching process.

The regular expression RE = (Start|Stop[ ]+)([0-9]+) defines two components:

- The strings "Start" and "Stop" followed by spaces — if matched, the string will be copied into the object Par:Command.
- The integer representing the device number — if matched, the number is copied into the object Par:DeviceNumber. In that case the string representing the device number will be automatically converted into integer according to the destination type (Par:DeviceNumber).

Any change of the Par:Input content results in a change of STR state. In the beginning it should be in the state INIT. After the first change it goes to MATCH or NOMATCH. With the Cmd:Set we activate the STR for the next operation — it goes to the state DEF. If we enter a proper string into the Par:Input object (e.g., "Start 5"), STR goes to the state MATCH. Note that the matching process is case sensitive and only strings that equal regular expression match.

To understand fully the functioning of the state machines used in tests we also need the project configuration. Below we show only the configuration of the STR string and the cooperating PAR objects:

```
//Configuration for Par object(s)
PAR Name = "Par:Command"
 Cat = "PP"
 Format = "string"
 Unit = " "
 InitValue = "0"
PAR Name = "Par:DeviceNumber"
 Cat = "PP"
 Format = "int"
 Unit = " "
 LimitLow = 0
 LimitHigh = 0
 InitValue = 0
PAR Name = "Par:Input"
 Cat = "PP"
 Format = "string"
 Unit = " "
 InitValue = ""

//Configuration for Str object(s)
STR Name = "Str:Command"
 Input = "Par:Input"
 RegularExpression = "(Start[ ]+|Stop[ ]+)([0-9]+)"
 Substring1 = "Par:Command"
 Substring2 = "Par:DeviceNumber"
```

Appendix Q

Other_Outputs Project

The Project

The project shown in Figure Q.1 contains:

- A state machine of a type Test_NO: Test_NO:01
- A state machine of a type Test_AL: Test_AL:01
- UNITs NO4 and DI8 to simulate outputs and inputs in SWLab

The State Machine Test_NO

This is a state machine whose behavior is described by the ST diagram in Figure Q.2. The ST tables in Figure Q.3, Figure Q.4, and Figure Q.5 contain the details of the state machine behavior.

The objects are represented by:

- Cmd: Cmd:Test_NO of a type CMD
- N01: No:01 of a type NO
- N02: No:02 of a type NO

The state machine demonstrates the possible use of the NO object. The state machine has 2 NO objects: two of them are after the start-up in the state OFF passing the value 0 to its output (their *Out Data* are PAR objects) and the third NO object passes to the output the value 20 as its *Out Data* is the TAB object (the index of the TAB object is initialized to 0, which selects the Par:01). Let us make a few exercises:

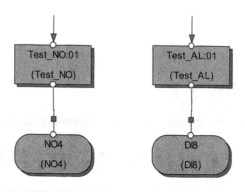

Figure Q.1 The Other_Outputs system.

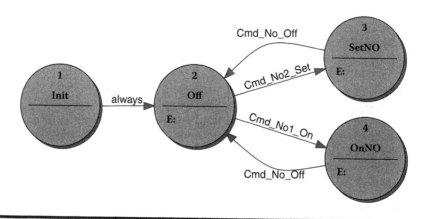

Figure Q.2 Test_NO: the ST diagram.

Off	Entry action	No1_Off No2_Off
	eXit action	
SetNO	Cmd_No2_Set	
OnNO	Cmd_No1_On	

Figure Q.3 Test_NO: the ST table of the state *Off*.

SetNO	Entry action	No2_Set
	eXit action	
Off	Cmd_No_Off	

Figure Q.4   Test_NO: the ST table of the state *SetNO*.

OnNO	Entry action	No1_On
	eXit action	
Off	Cmd_No_Off	

Figure Q.5   Test_NO: the ST table of the state *OnNO*.

First we send the command Cmd_No1_On: the state machine Test_No:01 goes to the state *OnNO* where it sends the command No1_On. The No1:01 object goes to the state CHANGED and from that moment if we change *Output (Par:01)* the No:01objects passes the new data to its output.

Receiving the command Cmd_No_Off the state machine Test_No:01 goes to the state *Off* where it sends the command No_Off to all NO objects which go to the state OFF passing the value 0 their outputs.

If the state machine Test_No:01 goes to the state *SetNO* it sends there the command Set to No:02 object, which passes the actual *Out Data* to its output. From this moment this value stays at the output until the command Set will be repeated, i.e., until the state machine leaves the state *SetNo* and returns to it.

Testing TAB Object

The project can be also used to demonstrate the use of the TAB object. The required objects are:

- Tab of a Type TAB
- Par:01, Par:02, Par:03, Par:04 of a type PAR
- No:3 of a type NO

We do not need any state machine for the test. We want only to show the typical use of the TAB object as a demultiplexer delivering to the output (linked to the NO object) several data values. After start-up the data of the NO object No:03, which uses Tab as *Out Data,* is set to the value 20 (Par:01) as the default index of the TAB object type is 0. Changing indexes of Tab from 0 to 3 we get the corresponding data of parameters: Par:01 through Par:04 to No:03 output.

To understand fully the test we need also the project configuration. Below we show only the configuration of the TAB configuration:

```
//Configuration for Tab object(s)
TAB Name = "Tab"
 Tab[0] = "Par:01"
 Tab[1] = "Par:02"
 Tab[2] = "Par:03"
 Tab[3] = "Par:04"
```

The State Machine Test_AL

This is a state machine whose behavior is described by the ST diagram in Figure Q.6. The ST tables in Figure Q.7, Figure Q.8, and Figure Q.9 contain the details of the state machine behavior. The test has been described in section *Example* in Chapter 13.

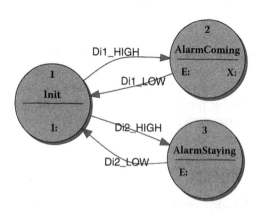

Figure Q.6 Test_AL: the ST diagram.

Init	Entry action	
	eXit action	
	Cmd_AI3	AI3_Staying
AlarmComing	Di1_HIGH	
AlarmStaying	Di2_HIGH	

Figure Q.7   Test_AL: the ST table of the state *Init*.

AlarmComing	Entry action	AI1_Coming
	eXit action	AI1_Going
Init	Di1_LOW	

Figure Q.8   Test_AL: the ST table of the state *AlarmComing*.

AlarmStaying	Entry action	AI2_Staying
	eXit action	
Init	Di2_LOW	

Figure Q.9   Test_AL: the ST table of the state *AlarmStaying*.

To understand fully the example we need also information from the configuration file shown below:

```
//Configuration for Al object(s)
AL Name = "Al:01"
  Cat = 1
  Text = "This is Al:01 alarm"
AL Name = "Al:02"
  Cat = 2
  Text = "No:01 has changed: %No:01"
```

```
AL Name = "Al:03"
 Cat = 4
 Text = "IDS_TEXT"

//Configuration for No object(s)
NO Name = "No:01"
 Format = "float"
 Unit = "mBar"
 ScaleMode = "Lin"
 ScaleFactor = 1
 Offset = 0
 OutData = "Par:01"
```

Appendix R

Counters Project

The Project

The project shown in Figure R.1 contains:

- A state machine of a type Test_CNT: Test_CNT
- A state machine of a type Test_ECNT: Test_ECNT
- A state machine of a type Test_UDC: Test_UDC:01
- UNITs: DI8:01 and DO8:01 to simulate inputs in SWLab

The State Machine Test_CNT

This is a state machine whose behavior is described by the ST diagram in Figure R.2. The ST tables are shown in Figure R.3, Figure R.4, Figure R.5, and Figure R.6.

After start-up the state machine waits in the state *Init* for initialization by Di HIGH, which starts the counter and forces the state machine to go to the state *Start*. The state machine never returns to the state *Init*.

To test the system we have to start SWLab and SWMon. The objects are represented by:

- Cnt:01 of a type CNT
- DiCountedByCnt of a type DI
- LED:Cnt_OVER of a type DO

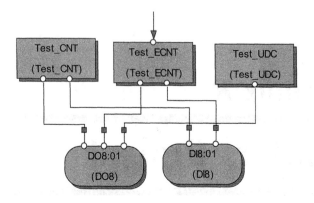

Figure R.1   The Counters system.

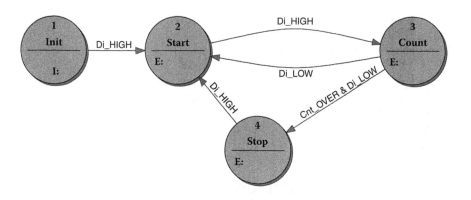

Figure R.2   Counters: the Test_CNT ST diagram.

Init	Entry action	
	eXit action	
	Di_HIGH	Cnt_Restart
Start	Di_HIGH	

Figure R.3   Test_CNT: the ST table of the state *Init*.

The CNT object has to count changes (to HIGH) of a digital input Di:CountedByCnt. After start-up the state machine is in the state *Init*. The first change of Di:CountedByCnt to HIGH starts the counter (in the Input

Start	Entry action	LED_Off
	eXit action	
Count	Di_HIGH	

Figure R.4   Test_CNT: the ST table of the state *Start*.

Count	Entry action	Cnt_Inc
	eXit action	
Stop	Cnt_OVER & Di_LOW	
Start	Di_LOW	

Figure R.5   Test_CNT: the ST table of the state *Count*.

Stop	Entry action	LED_On Cnt_Restart
	eXit action	
Start	Di_HIGH	

Figure R.6   Test_CNT: the ST table of the state *Stop*.

Action) and changes the state to *Start*. On entering the state *Start* the output LED:Cnt_OVER is switched off (at that first entry it has been initialized to off). The state machine continues the state changes (Di is HIGH) going to the state *Count* where it sends the Inc command to the counter. On Di LOW it returns to the state *Start*. The next Di HIGH means a transition to the state *Count* and the counter increments and so on. This switching between these two states last until the counter reaches the *expiration value* and goes to its state OVER. The counter OVER causes a transition to the state *Stop* where LED:Cnt_OVER is switched on signaling the expiration and the counter is restarted. The next Di HIGH causes a transition to the state *Start* and the counting will be repeated.

We have to be careful in specifying the transition in the state *Count*: the transition condition Cnt_OVER to the state *Stop* would not work. It will cause an immediate transition to *Count* if Di is HIGH (which generates *Cnt_OVER*). This in turn will cause a further transition to the state *Start* and the state machine will loop through the states: *Start–Count–Stop*. If we use the Cnt_OVER and Di_LOW condition the transition to the state *Stop* is delayed until Di is LOW. We note also that the sequence of Transition expressions in the table is relevant.

The State Machine Test_ECNT

This is a combinational control system whose behavior is described by the table *Always* in Figure R.7.

The Relevant objects are represented by:

- Test_ECNT:Cmd of a type CMD
- Ecnt:01of a type ECNT
- DiCountedByEcnt of a type DI
- LED:Ecnt_OVER of a type DO

To understand the test we need also information from the configuration file below:

```
//Configuration for Ecnt object(s)
ECNT Name = "Ecnt:01"
  Const = 5
  Input = "Di:CountedByEcnt"
  UpValue = 1
```

Always	Cmd_RestartEcnt	LED_Off Ecnt_Restart
Always	Ecnt_OVER	LED_On MyCmd_Clear

Figure R.7   Test_ECNT state machine: the table *Always*.

The State Machine Test_UDC

This is a state machine whose behavior is described by the ST diagram in Figure R.8. The control is specified mainly in the table *Always* (see Figure R.9). The state *Start* (Figure R.10) is used only to enable the switchpoints.

To test the system we have to start SWLab and SWMon. The objects used by the state machine are represented by:

- Swip_for_UDC:01 of a type SWIP
- Swip_for_UDC:02 of a type SWIP
- LED:Udc_5 of a type DO
- LED:Udc_InRange of a type DO

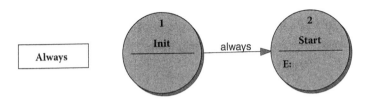

Figure R.8 Test_UDC: the ST diagram.

Always	Swip_Five	Do_High
Always	Swip_LOW | Swip_HIGH	Do_Low
Always	Swip2_IN	Do2_High
Always	Swip2_LOW | Swip2_HIGH	Do2_Low

Figure R.9 Test_UDC: the table *Always*.

Start	Entry action	Swip_On Swip2_On
	eXit action	

Figure R.10 Test_UDC: the ST table of the state *Start*.

Indirectly the state machine is linked with objects:

- Udc:01 of a type UDC
- Par:UDCLimiLow of a type PAR
- Par:UDCLimiHigh of a type PAR

As we have not provided any loading of PP parameters after start-up for both PAR objects that define the SWIP limits stay UNDEF. This in turn causes the Swip_for_UDC:02 to switch also to the state UNDEF. The supervision of the Udc:01 counter is effectively disabled until the parameters are set to a new value.

To understand the test we also need information from the configuration file below:

```
//Configuration for Swip object(s)
SWIP Name = "Swip_for_UDC:01"
 Input = "Udc:01"
 LimitLow = 5
 LimitHigh = 5
SWIP Name = "Swip_for_UDC:02"
 Input = "Udc:01"
 LimitLow = "Par:UDCLimitLow"
 LimitHigh = "Par:UDCLimitHigh"

//Configuration for Udc object(s)
UDC Name = "Udc:01"
 Unit = "<none>"
 UpInput = "Test_CNT"
 UpValue = 4
 DownInput = "Di:UdcDownInput"
 DownValue = 1
 ClearInput = "Di:UdcClearInput"
 ClearValue = 0
```

Appendix S

Attributes of RTDB Objects

		VFSM	CMD	TI	AL	DI	DO	NO	NI	SWIP	XDA	PAR	OFUN	STR	CNT	DAT	UNIT	ECNT	UDC	TAB
	IAtt_None	R	R	M	R	R	M	M	R	M	M	R	M	M	M	R	R	M	M	M
.Val	IAtt_Value	R	R	M	R	R	M	M	R	M	M	R	M	M	M	R	-	M	M	M
.SvM	IAtt_ServiceMode	M	M	-	-	M	M	-	-	M	-			-	-	-	-	-	-	-
.SvV	IAtt_ServiceValue	M	M	-	-	M	M	-	-	M	-			-	-	-	-	-	-	-
.PeV	IAtt_PeripheralValue	R	M	-	-	R	R	-	-	R	-			-	-	-	-	-	-	-
.VI	IAtt_VI	R	-	-	-	-	-	-	-	-	-			-	-	-	-	-	-	-
.StN	IAtt_StateName	R	-	R	R	-	-	R	R	R	-	R	-	R	R	R	-	R	R	-
.AIL	IAtt_AssocItemList	R	-	-	R	-	-	-	-	-	-	-	-	-	-	-	-	R	-	-
.Typ	IAtt_TypeName	R	-	-	R	-	-	-	-	-	-	-	-	-	-	-	-	R	-	-
.CnC	IAtt_CountConstant	-	-	M	-	-	-	-	-	-	-	-	-	-	M	-	-	M	-	-
.CnR	IAtt_CountRegister	-	-	R	-	-	-	-	-	-	-	-	-	-	R	-	-	R	-	-
.Cat	IAtt_Category	-	-	-	R	-	-	-	-	-	-	R	-	-	-	-	-	-	-	-
.Frm	IAtt_Format	-	-	-	-	-	-	R	R	-	-	R	-	-	-	R	-	-	R	-
.Uni	IAtt_PhysicalUnit	-	-	R	-	-	-	R	R	-	-	R	-	-	-	R	-	-	R	-
.LiL	IAtt_LimitLow	-	-	-	-	-	-	-	R	M	-	R	-	-	-	-	-	-	-	-
.LiH	IAtt_LimitHigh	-	-	-	-	-	-	-	-	M	-	R	-	-	-	-	-	-	-	-

		VFSM	CMD	TI	AL	DI	DO	NO	NI	SWIP	XDA	PAR	OFUN	STR	CNT	DAT	UNIT	ECNT	UDC	TAB
.IVa	IAtt_InitValue	-	-	-	-	-	-	-	-	-	-	R	-	M	-	-	-	-	-	-
.Dat	IAtt_DataValue	-	-	-	-	-	-	R	R	R	-	M	-	-	-	M	-	-	M	-
.Txt	IAtt_Text	-	-	-	R	-	-	-	-	-	-	-	-	-	-	-	-	-	-	-
.Ack	IAtt_Acknowledge	-	-	-	W	-	-	-	-	-	-	-	-	-	-	-	-	-	-	-
.Tim	IAtt_Time	-	-	-	R	-	-	-	-	-	-	-	-	-	-	-	-	-	-	-
.ScF	IAtt_ScaleFactor	-	-	-	-	-	-	R	R	-	-	-	-	-	-	-	-	-	-	-
.Ofs	IAtt_Offset	-	-	-	-	-	-	R	R	-	-	-	-	-	-	-	-	-	-	-
.ScM	IAtt_ScaleMode	-	-	-	-	-	-	R	R	-	-	-	-	-	-	-	-	-	-	-
.Lst	IAtt_List	R	R	-	-	-	-	-	-	-	-	-	-	-	-	-	-	-	-	-
.PAd	IAtt_PhysAddr	-	-	-	-	-	-	-	-	-	-	-	-	-	-	-	R	-	-	-
.Com	IAtt_CommPort	-	-	-	-	-	-	-	-	-	-	-	-	-	-	-	R	-	-	-
.Trc	IAtt_Trace	M	M	M	M	M	M	M	M	M	M	M	M	M	M	M	M	-	M	M
.RMo	IAtt_RunMode	M	-	-	-	-	-	-	-	-	-	-	-	-	-	-	-	-	-	-
.NSt	IAtt_NextStep	R	-	-	-	-	-	-	-	-	-	-	-	-	-	-	-	-	-	-

Note: - = none, R = read only, M = read/write, W = write only.

Appendix T

StateWORKS Tools and Components

StateWORKS Studio

StateWORKS Studio (SWStudio) is an Integrated Development Environment that is used to specify and test single state machines and systems of state machines. SWStudio consists of two parts: SWEdit and Project Manager. SWEdit is a state machine editor to create and modify state machines, UNIT types, and string resources. Project Manager is used to specify the RTDB (detailed description can be found in the manual[1]). An introduction to the use of StateWORKS development tools is in the user's guide.[2] The Web site[3] is also a good source of information about all aspects of StateWORKS development environment and runtime system.

State Machine Specification

For a state machine specification a state transition (ST) diagram and ST tables are used. The specification is done in a virtual environment, which is defined by RTDB objects foreseen for the state machine. The definitions of the required objects are placed in the I/O Object Dictionary. The bases of the specification are Input Name Dictionary and Output Name Dictionary defined on values of RTDB objects. The required states are defined in the State Name Dictionary. Using those Dictionaries, states and their transitions and actions are defined. The result of the state machine specification is a new VFSM object type.

UNIT Specification

UNITs are defined in a table. The specification is done using an RTDB UNIT-object type foreseen as interface between the RTDB and I/O Handlers or Output functions. The result of the specification is a new UNIT object type.

String Resource Specification

String resources are replacements for Windows resource IDs, which are used for internationalization. The specification is done in a table (IDS and corresponding string). The result of the specification is used by a StateWORKS runtime system for internationalization of text alarms.

Definition of Object Properties

Objects are created and their properties are specified in Project Manager. The specified objects define the RTDB.

Definition of System of State Machines

By specification of VFSM objects, properties links (command and state) are established. Those links define the system of state machines. The system can be displayed in the State Machines System (SMS) diagram.

Building

By Building the specification results are transformed into several files located in the destination folders:

- Configuration: RTDB configuration (SWD), state machine and UNIT definition (IOD and H), state machine behavior (STR), string resources (SRC and resource.h), system specification for embedded application (CPP)
- XML: state machine descriptions (XML) and entire project description (PRJ.XML)
- Graphics: diagrams and tables of all state machines and a system diagram (JPG or WMF)

Building of incomplete systems is possible.

Testing

The system can be tested at any phase of the specification (i.e., when incompletely specified) using Tools commands to start: SWLab, SWMon, SWTerm, and SWQuick. The logging files SULOG.TXT and TRACE.TXT are stored in the Configuration folder.

StateWORKS Simulation

SWLab is an RTDB-based Windows application whose user interface contains typical inputs and outputs used in industrial control:

■ Eight switches to simulate digital inputs
■ Eight LED to simulate digital outputs
■ Four slides to simulate analog inputs
■ Four gauges to simulate analog outputs

The SWLab inputs and outputs are available if the system configuration contains standard UNITs: DI8, DO8, NI4, and NO4.

SWLab can also be used to test systems without simulated input/outputs: all objects are accessible in monitors.

At SWLab start a SULOG.TXT file is created, which contains information about missing or erroneous objects and their properties.

StateWORKS Monitors

SWStudio contains three different monitors.

■ SWMon is the most powerful monitor, which allows watching and manipulating several RTDB objects at the same time. All attributes are displayed and can be changed if writable. Several object views are provided: VFSM or UNIT related as well as user defined. Trace mode and Debug mode are at hand.
■ SWTerm is a terminal monitor. All activities (commands and system answers) are logged. Commands can be written on terminal console or taken from a command file. That monitor is used for repeated testing with command files.
■ SWQuick is a simple monitor used to access a single object at a time. The monitor is very handy and gives a quick overlook of all object properties.

StateWORKS Runtime Systems

The RTDB can be used to build control applications. SWLab is an example of a standard runtime system, used for testing. The list of standard runtime systems includes:

- WinStExec: an RTDB-based Windows application with built-in standard I/O-Handler (with DLL interface) plus some specific I/O-Handlers (Serial and USB).
- StExec: a variant of WinStExec but terminal based.
- LinuxExec: a variant of WinStExec running under a Linux operating system.
- StateWORKS can also be used in disk-less RTOS environments.

Recommended Reading

1. SW Software: StateWORKS. Reference Manual for the Class Library.
2. SW Software: StateWORKS Development Tools. User's Guide.
3. www.stateworks.com.

Index